Sonic UX

Sonic UX

The Science of Sound in User Experience Research

KC Collins

foreword by Andrew Scott Duncan and Katie Edmonds

The MIT Press
Cambridge, Massachusetts
London, England

The MIT Press
Massachusetts Institute of Technology
77 Massachusetts Avenue
Cambridge, MA 02139
mitpress.mit.edu

The MIT Press would like to thank the anonymous peer reviewers who provided comments on drafts of this book. The generous work of academic experts is essential for establishing the authority and quality of our publications. We acknowledge with gratitude the contributions of these otherwise uncredited readers.

This book was set in Stone Serif and Stone Sans by Westchester Publishing Services.
Printed and bound in the United States of America

Library of Congress Cataloging-in-Publication Data is available.

ISBN: 978-0-262-05373-0

10 9 8 7 6 5 4 3 2 1

EU Authorised Representative: Easy Access System Europe, Mustamäe tee 50, 10621 Tallinn, Estonia | Email: gpsr.requests@easproject.com

Contents

Foreword: Sound Is the Secret

Once upon a time humans interfaced with computers through command lines, and the computational capacity of computers was the limit of their usefulness. Their screens were green and their sounds were mostly mechanical save for a few simple sine wave beeps. As their processing increased, new interfaces were developed. Alan Cooper's Visual Basic launched the creation of graphical user interfaces for screens, and peripheral features were added to enable networking, image capture, global positioning systems, microphones, and speakers. These machines were engineered to fit in homes, in cars, in electrical grids, and anywhere else they might add efficiency, productivity, or profit. Experimentation with techniques for interacting with them exploded in countless directions, and digital interfaces slowly crept upstream from an engineering afterthought to the field we now know as user interface and experience design. The role of sound in this ecosystem was an afterthought of an afterthought. This book is a leap forward into a new paradigm of experience design that could accomplish what experience designers have been trying to accomplish for years: the evolution of interface-centered design to experience-centered design.

Sound wasn't absent from the early days of computers, though it was on a separate developmental path from interface and experience design. As computers were advancing in their capacity and application, sound-oriented experimentation tended to focus on using computers to create new kinds of sounds. Composition and engineering were the drivers of innovation in this space. From Fourier's investigation of waves at the end of the industrial revolution to frequency modulation and John Chowning's synthesis algorithm that enabled the Yamaha synthesizer explosion in the 1970s and 1980s, methods and tools for creating sound didn't yet apply sound as a means for interacting with the products or the built environment. Once digital devices became commonplace, and users without technical expertise started to demand more usable products, sounds became features of the products we used every day. However, sound experts had yet to be considered essential contributors to these purely functional features.

All this started to change with the birth of the mobile and tablet market. Reading this book sparked some deep reflection on our own journey into sound for experience design that began at the birth of smartphones and tablets. The first time we collaborated on building soundful digital experiences was back in 2010 when Apple announced the iPad. A small group formed around the idea of adapting graphic novels to this new format. Sound design was never scoped to be part of the production process, but as the team finalized decisions about the interface and started building out a minimum viable product, the lack of sound design became glaring. Stories need sound, so in the classic Silicon Valley style of talking to friends of friends while bar hopping in the Mission district, the developers connected with a sound designer who was willing to take on these challenges, which were going to require some new thinking and experimentation. Working without a compass or map, and after overcoming a few existential threats, we produced a truly engaging story experience. The book you're reading now is a book we wished we had all those years ago.

Looking back after years of working in this field, those struggles feel small because they were the size of an individual handheld screen. In the years since, we've worked hard to expand the value proposition of interface design and transform it into an area of expertise that creates coherent experiences across multiple touchpoints in a user's journey. Despite this effort we find that most executives and investors are still looking at screens in the same way we were back in the early days of mobile. This book is a wide path out of that thinking, and what makes it especially interesting is that the path is not forged by graphic design or industrial design. It's forged by sound design.

As you endeavor to become an expert in sound design for experiences, you're going to come up against some challenges that will require more of you than just technical chops. Not everyone is going to understand what you're doing or the value that you bring to the end user. You are most certainly going to experience a significant headwind from skeptical leaders who have decision-making power over schedules and budgets. There will be people who think sound is a nice trick, a pretty addition, while others will think it's a waste of money. Consider that success is not just completing a piece of work. Success is also winning the debate on the value of your work and enabling these designs to get into the market. It's rare that your work is so good that it doesn't need evangelizing.

You have more in common with the founders of experience design than you have with contemporary experience designers, and that makes this an exciting time. Sound travels beyond screens, and emerging fields like service design argue that every human interaction required to deliver value to an end user can be designed irrespective of the presence of a screen. However, as interface design is slowly pulled back under the chief technology officer's control due to the development of artificial intelligence (AI) tools for engineering the front end of products, creative skills can be valued less than technical skills. Sound designers always have and always will think experientially, and the value sound design offers is not an addendum or a peripheral to user-centered experiences if it's a fundamental partner in value creation. Are your client's product/service ecosystems legitimately user-centered? If so, are they truly experiential? Do they use sound to make their offerings experiential? These are questions that can change the conversation if they're quickly followed by proofs of concept that are well made by thoughtful creatives. The good news is that you no longer have to learn as you go, because this book is your guide.

These are early days in this field, and sound can grow to be a part of every company's relationship with its customers. We believe that branding, packaging, and product designers are sound designer's collaborators because they know how to build design systems and experiences. The current hurdles for a sound designer are more related to the "how" of expressing an audio design system. Customers know that a clean interface or beautiful environment is the bare minimum for an elevated experience. They are looking for more, but are getting more of the same. Transitions between digital and physical experiences are still jarring. Experience design in its current state has yet to leverage all the sensory opportunities that deliver seamless journeys across touchpoints. Sound can be the glue between physical products, digital products, and service experience. As a sound designer for user experiences, these interactions with products are firmly within your purview. To leave these decisions to other designers, or to engineers, is to leave work on the table.

As you navigate the world, be receptive to the opportunities for sound to play a part in the world you interact with, and embrace the creative challenge of incorporating sound into any experience. If something is designed, it can be redesigned. If something can be designed, it can be designed with sound. The discourse around creating great product and service experiences is dense with presumed panacea. Human-centered design and design thinking, behavioral economics, and big data, not to mention AI, claim to provide conclusive value to the end user, yet fail to do so. When teams include sound in the core value proposition of experience design, great breakthroughs are possible. The effort to argue for this has set the stage for the comprehensive guide that this book delivers. It's a technically and conceptually usable reference for bringing this work to market. We think you should approach this text with the same eagerness you would books written by experts like Don Norman or Steve Krug. This book will live on our desks, not on our shelves, for a long time to come.

Andrew Scott Duncan and Katie Edmonds
ASDSoundDesign.com and katiedmonds.com

Introduction

I open my car door, sit down, and close the door with a satisfying *ka-thunk*. My phone chimes at me to signal an incoming text message, and I pull it out and quickly type out a reply, the onscreen keyboard sounds clattering away. I stick the car key in the ignition and an annoying beep reminds me to put on my seat belt. I click the seat belt in and turn on my music as the GPS cuts in to tell me where to turn next.

We're surrounded by designed product sounds throughout our days. We will typically hear them many times in any given day, but how many of us recognize the role that these sounds play in our lives, or the work that went into selecting and crafting these sounds? If we take a moment to think through the purpose of these sounds, we might come up with a list like this:

Car door: The purpose of the car door's sound is to signify solid craftsmanship and give a sense of weight and safety to the door. Car companies have large teams of sound designers and engineers who will do things like add weights inside car doors to make sure they deliver the right feeling associated with the car. There are emotional associations we might have with that sound, too, depending on our connection to the car or its sonic uniqueness (we might love that creak on our old Volkswagen, for instance, and associate it with the summer trips we took). There may also be brand associations with the sound: an expensive luxury vehicle must sound more solid and effective than a cheap car.

Seat belt click: As with the car door, the click has been designed to be satisfying. It also helps to signify that the seat belt has been connected properly. If we don't hear the click, whether we notice it consciously or not, we may keep fiddling with it thinking that it hasn't worked.

Phone chimes: The purpose of our notification sounds on our phones is to indicate an incoming message. There may also be emotional associations we have with the sound that becomes Pavlovian (we may have a positive response if we get a lot of messages from friends, or a negative response if we're pestered by many work messages). Just as Pavlov conditioned animals to respond to sounds by associating them with food, our phone pinging at us causes a physiological response of excitement (Fortin et al., 2019). Alternatively, we may develop what David Laramie termed *ringxiety*, in which our response is triggered by someone else's phone or a similar sound (see Frissen, 2020). There may also be branding associations with the sound (an iPhone has different default sound settings than an Android phone, for instance), or if we've personalized some chimes we may have certain feelings and emotions associated with the particular chime we hear.

Keyboard sounds: Many people turn off the keyboard sounds, and some young people think keyboard sounds are a signifier that you're old: *Slate* magazine published an article entitled, "This holiday season, talk to your baby boomer parents about their annoying cellphone noises," in which they ask, "Surprisingly, no one has been murdered in public based on their decision to keep 'keyboard click' sounds turned on as they type. But do you really want to risk being first victim?"—see Graham 2015). Many people prefer the sonic feedback they get when they have lost the sense of touch feedback, especially if we've been conditioned to expect them, as older users have.

Seat belt alarm beep: Here's a sound that is designed to annoy us in order to get us to take an action. We are rewarded with the blissful absence of annoyance when we plug in the seat belt. The sound serves an informative and behavioral (negative reinforcement) role, in other words, and in this case, must comply with government regulations.

I've just shown how sounds are designed to serve multiple purposes, often at the same time. Yet we rarely pay any attention to how and why sounds are designed, even though they influence our feelings, associations, moods, and behaviors. This book aims to change that lack of attention by teaching you how you can leverage the power of sound for your products. A lot of product sound design literature is centered around how to avoid annoying people (e.g., Drescher, 2008), and we'll cover that too, but more importantly, we'll look at how to use and test sound to improve your products, strengthen your brand, and deliver the user experience that your users deserve.

We are an *ocularcentric* society, which is to say, we center vision, rather than other senses. But our capacity for sound is greater than vision in many ways: the frequency range of our hearing (20 hertz [Hz] to 20k Hz) is more than 400 times broader than our visual spectrum (405 terahertz [THz] to 290 THz). We can discern about 150 colors, but more than 1,300 tones (Rychtáriková et al., 2010). We are able to locate sounds in three-dimensional space allowing for 360 degrees of directionality and envelopment, whereas our field of vision is comparatively low (roughly 200 degrees, or 120 degrees with both eyes, and 40 degrees each side for peripheral vision). Despite our capacity for sound, then, sound design is surprisingly under-leveraged in our product design, and there is so much opportunity for innovation in this space.

0.1 What Is Sound Design?

What is sound design, exactly? How do we *design* a sound? For the purposes of this book, we will think of sound design as the process of creating one or more sounds through recording, selecting, editing, and manipulating those sounds in a product. A product could be a physical object, like a car or microwave oven, or it could be a piece of software or video game. For our purposes it could even be a service that is either physical or virtual.

For a long time, product sound design was limited to the desire to dampen or eliminate unwanted sounds from a product, such as typewriters, vacuum cleaners, and pretty much anything with a motor. It's interesting that such electro-mechanical sounds have become a part of digital sound design: The *skeuomorphic* representation of the sounds of analog or mechanical movements into the digital realm is found all over sound design today, from the click-whirr of a digital photograph being taken to the aforementioned typewriter keys on a touchscreen.

The term *sound design* is borrowed from film sound. The process of sound design for film shares a lot of commonalities with sound design for products: Both are about driving a (often subconscious) response from the listener, and both involve a careful crafting of the sound to create that response. The main difference between the two forms of sound design, however, is the role of *user experience research* in product design: the subject of this book.

The use and/or placement of sound in the product (i.e., deciding when and where a sound plays) is one key consideration of the sound designer, as is deciding which functions a sound should play (such as enhancing usability, driving emotion, or reinforcing branding). Another key consideration is how to manipulate sounds to get the right sound in the right place. When it comes to product sound design, then, we can define the sound designer's role as getting *the right sound in the right place at the right time with the right impact on the user*. However, this definition comes with caveats, as we will see later.

When it comes to products, there are several types of sounds to consider:

Intentionally designed sounds: These are sounds that have been put in the product to enhance the user experience, branding, or functionality. The sounds are chosen and/or designed by a designer, and typically provide a communicative function. These include, for instance, interface sounds, start-up sounds, sonic logos, warnings or alerts, and so on.

Incidental sounds: These sounds occur through interacting with or in the operation of the product. These may be sounds created by motors, servos, and gears in the device; by the function of the product (e.g., the vacuum cleaner's suction, the coffee maker's grinding); or by other causes. These are sometimes designed by a sound designer working with an engineer and, as will be shown, can have a considerable impact on your users, so *should* be designed carefully.

Packaging sounds: For physical objects, the sensory experience of unboxing also plays a role in product design, and the sound of packaging should also be considered. As will be shown, packaging sounds can also influence the perception of your product.

Contextual sounds: Will the product be purchased in a physical store? Where will the product be used? How will the product be influenced by the sounds around it, and how will the product's sounds interact with the people (users and non-users) in its environment?

Product sound design, as I talk about in this book, refers to the intentional designing of sound in products to enhance user experience. Product sound requirements can vary depending on individual product, brand/company products more generally, regulatory requirements (noise emission standards, consumer health and safety regulations around volume, electromagnetic and radio frequency interference), accessibility requirements, industry standards such as loudness (e.g., see Duggal, 2024), cultural and social considerations, privacy laws, competitor assessments, technical restrictions, and other reasons. There is a lot to consider when designing sound for a product.

0.2 Stop Ignoring Sound! Top Ten Functions of Product Sound Design

It should now be clear that sound design can bring a lot to the table when it comes to products. Sound is more than just pretty aural wallpaper that can be left as an afterthought. I'll be discussing many of these functions in more detail throughout the book, but for now, let's examine just some of the functions that sound plays in products.

0.2.1 Sound as Communication

Perhaps the most important role of sound is to communicate. Whether it's birds singing to defend territory, people discussing a movie, a dog barking to warn you to stay away, or any other sound, a lot of intentionally produced sounds in nature are communication. When it comes to sound design in products, communication is still the most important role: Sound is used as a notification that something is happening, is about to happen, or has happened. It serves as feedback that we took an action, as a warning to stop taking an action, or as a warning that we need to take an action. It draws our attention to an event, telling us that there is something we should pay attention to. My dryer plays a little jingle when the clothes are dry. My car warns me that I need to buckle my seat belt. The early morning delivery truck next door warns me that the truck is backing up with incessant beeping. All of these sounds contribute to the basic usability of our products.

As Donald Norman (2009, p. 59) notes in *The Design of Future Things*, "sounds are an automatic result whenever objects move, whenever they meet one another, scraping, colliding, pushing, or resisting. Sounds tell us where things are located in space, but they can also reveal their composition (leaves, branches, metal, wood, glass) and activity (falling, sliding, breaking, closing) as well. Even stationary objects contribute to our aural experience, for the way that sounds are reflected and shaped by environmental structures gives us a sense of space and our location within it." Thus, a lot of information can be packed into a single sound.

0.2.2 Sound to Reduce Cognitive Load

Because sound provides information beyond visuals, sound can reduce cognitive load (in simplified terms, your working memory resources), making complex tasks easier. If information across our senses conflicts with what we're trying to take in, this can make cognitive load higher. For instance, if you've ever turned off the radio in the car in a busy city because you're trying to concentrate on avoiding pedestrians or trying find the right street, you were trying to lower your cognitive load by reducing the stimuli your brain has to take in. If we have information that is useful across different modalities, we can take in more information. For instance, if we hear the GPS voice speaking to us while we view those same streets, it's easier for us to navigate because our eyes can then scan for road hazards instead of trying to read a map and street signs. When users are consistently distracted by all the visual clutter vying for their attention in their lives, isn't the addition of sound something to consider?

0.2.3 Sound to Change Behavior

Sounds can be associated with specific conditions in a very Pavlovian way, through our associations with those sounds. For instance, loud growly sounds are associated with power, and Harley Davidson capitalized on this association with their motorcycle engine sounds. As we move toward electric motorcycles, it will be interesting to see if these sounds remain in Harleys even when engines are no longer making much noise (I suspect, as with electric cars and their muscle car mods, they will). Sound can also be used to change a user's behavior, using positive or negative reinforcement. The incessant beeping of the seat belt sound in the car, for instance, is useful as a negative reinforcer to prompt me to fasten my belt in order to get the sound to stop.

0.2.4 Sound to Drive Emotional Response

Sound can alter the way we feel about a product, making us feel happy, sad, delighted, and so on. Like great visual design, sound can add to the overall perception of a product. We all know the feeling of opening a beautifully designed box and seeing a pristine, beautiful new product inside. But this effect also extends to sound, and having that perceptual delight of hearing a great sound in an interaction can also fulfil our aesthetic pleasure. As a child, I loved the sound of Mario jumping in Nintendo's *Super Mario*

Bros games (1985 onward), and would press the button over and over to hear that sound. For me, the sound became associated with the anticipatory excitement of playing a game.

0.2.5 Sound for Brand Identity and Recognition

Just as visual elements like logos and colors contribute to brand recognition, sound can also play a role in establishing a brand's identity. Iconic startup sounds, jingles, or notification sounds can become strongly associated with a brand. Sound can associate a product with the brand's attributes, be a brand differentiator, and help to establish a unique brand personality. The sound of a Mac booting up is different from a Windows machine, and signals brand choice not just to the user but to everyone around them. Vehicles similarly sometimes have unique sounds (Harley Davidson fought a legal battle to try to trademark their v-twin sound, for instance—see Flynn 2022), and today many global companies are working to establish their unique sonic brand.

0.2.6 Sound to Extend and/or Enhance Engagement

Sound can make interactions more engaging and interactive. In video games, for example, sound effects and music play a significant role in immersing players in the virtual world. Immersive sound design encourages users to engage more deeply with a product, enhancing their overall involvement and enjoyment. High-quality sound design can contribute to the perception of a product's overall quality. Satisfying and well-designed sounds, such as the subtle click of premium buttons on a device, can enhance the perception of craftsmanship and attention to detail. Thoughtful and unexpected sound interactions can surprise and delight users. Playful sounds triggered by certain actions, like Easter eggs, when the user discovers a secret functionality, can create a sense of fun and wonder.

0.2.7 Sound for Usability

Sound design can enhance the usability of products, highlighting functionality and memorability. Sound cues can make interfaces more user-friendly by indicating when certain actions are available or complete. For instance, a gentle chime might signal that a button has successfully been clicked and a form submitted. Auditory cues can guide users through interfaces and interactions. Consistent sound cues throughout interactions with a product create a familiar and predictable environment, making it easier for users to understand and navigate the product. Well-designed sound cues provide clear and distinct feedback, making interactions more intuitive. Providing options for users to customize sound settings allows individuals with varying preferences and sensitivities to tailor the experience to their needs. By considering usability and accessibility in sound design, product designers can create products that are user-friendly, inclusive, and capable of accommodating a wider range of users, regardless of their abilities.

0.2.8 Sound for Aesthetics and User Experience

Just as visual design affects the look and feel of a product, sound design contributes to the overall aesthetic and user experience. The aesthetic-usability effect is well studied in terms of visual design (i.e., a beautiful product is felt to be more functional), but sound can play a role in aesthetics and perceived usability as a result as well. Pleasant and well-thought-out sounds can enhance the perceived quality of a product. The choice of sound effects, music, and ambient sounds can help set the tone for the product, whether it's a mobile app, a video game, or a physical device.

0.2.9 Sound to Anthropomorphize

Sounds can help to anthropomorphize an object, in order to create a sense of attachment, give life to an object, and build an emotional connection. The first fifteen or twenty minutes of Pixar's movie *Wall-E* (2008) don't have much dialogue in the traditional sense of using words, but are carried through the robot's sounds (much as R2D2 in *Star Wars* (1977) was able to communicate by beeping at us). The robots in *Wall-E* are given personality and feelings—they are given character and *life*—through their sound design. As an experiment I added sounds to a virtual parking meter and tested the impact on anthropomorphism—and found that, indeed, sound increased the level of anthropomorphism (Collins et al., 2025).

0.2.10 Sound for Accessibility and Inclusion

For users with visual impairments, sound provides an alternative form of feedback. Auditory cues and feedback can help people understand the status of a product, navigate interfaces, and complete tasks. Sound can work in tandem with assistive technologies like screen readers to provide a more comprehensive experience for users with disabilities. Sound can warn all users about potential hazards or important notifications. Sound can also serve an accessibility function when our eyes may be otherwise engaged (such as while driving a car) or when we cannot use our eyes (such as in the dark or in a smoke-filled environment).

Having just gone through a list of functions or roles sound can play in your product design, isn't it clear that thinking through those functions should be done with some careful consideration and iterations on the design, instead of at the last minute? I hear time and again from sound designers that they've been dropped into a project at the very last minute—sometimes the last day before a product is about to be launched—and expected to "just add some sounds." I hope that this book will show that getting the right sound can be critical to the user experience and to the brand, and that this process takes time and iterative practice, just as with other aspects of design. This means that the sound design team should have a seat at the table early on in the design process, and should be able to contribute ideas and give input on how sound might enhance or contribute to the product.

It takes time to get it right: lots of time! Microsoft's Conor O'Sullivan, for instance, tells how it might take months to get a single sound right, and a sound may go through a hundred cycles of listening and iterating (Novet, 2021). You wouldn't leave the visual design of a product until the last minute, so why would you leave the sound? Planning and designing with sound in mind can save time and money, and create a more memorable user experience.

0.3 What Is Sonic User Experience?

In this book, I use the term *user experience* to describe the overall experience that a user—or consumer—has with a product. This term centers the user's needs and desires in the design, as critical to the entire iterative process of design. User experience (UX) includes how much the users are attracted to and engaged with a product, the ease of use (i.e., the usability), the branding, and the pleasure or other feelings associated with using a product. UX covers the overall quality of a person's interaction with a product or service, and how that interaction shapes their perceptions, emotions, and satisfaction. It encompasses all aspects of the user's journey, from their initial awareness and discovery of the product to their ongoing usage and long-term loyalty.

User experience research focuses on gaining a comprehensive understanding of these perceptions, emotions, and overall experiences when engaging with a product or service. It takes a holistic approach, considering various aspects such as user motivations, goals, and expectations throughout their experience with the product. UX research aims to uncover deeper insights into user behavior, preferences, and pain points, allowing designers to create more meaningful and emotionally resonant experiences. It often involves research methods such as interviews, surveys, observation, and user testing to gather data that inform the overall design strategy. The insights gathered from UX research inform design decisions, help prioritize features, and guide refinements throughout the product lifecycle. By involving users in the design process and understanding their needs, organizations can create products that are user-friendly, intuitive, and aligned with user expectations, leading to enhanced user satisfaction and engagement.

User experience design and research is a rapidly growing area as companies realize that products that are UX tested end up saving money in the long run, and the products result in better user engagement and sales. When it comes to the sounds described above, for instance, imagine if my car door sound sounded weak and tinny: Users might say the car *feels cheap*, although they may not be able to put their finger on *why* it feels cheap. Through gathering such data and an iterative design process, we might learn how to give that car the right sound and, as a result, the right feel for our users, encouraging them to choose our car over those of the competitors.

When it comes to user experience research applied to sound, there are additional considerations beyond standard UX testing, so this book aims to provide a basis on which to build this burgeoning field. Until quite recently, sound designers have rarely had to contend with the user experience process, since as described, sounds are often left until the end of the design cycle when it's too late to do much about them. Sound designers have learned which sounds work through intuition rather than data. While intuition can take us a long way, intuition can't provide the data we need to convince stakeholders we're on the right track and to invest further in sound, or to show how the sounds chosen are on-brand and provide the right experience for customers. Moreover, UX researchers may treat sound much like any other aspect of the product, and not understand some of the challenges of testing specifically for sound, which I'll cover here. Sound has some unique difficulties, and UX researchers should be aware of the particular needs when it comes to testing sound. In the chapters to come, we will be exploring many examples of what questions can be asked and how they can be answered during user experience research.

0.4 Sound and User Experience

A Reddit thread presents an interesting study in user responses to sound in general, but more specifically it was initially about the Sony PlayStation start-up sound.[1] The variety of responses are quite telling of why user studies should be run. Sony PlayStation had a beeping sound upon booting up, and by the time the PS5 came out, the sound was deemed to be loud enough to wake up nearby sleeping people, angering users. The firmware update that allowed users to turn off the sound was well received by users. The Reddit thread explaining how to turn off the sound received thousands of upvotes and was called "No more waking the wife." Interestingly, the highest upvoted comment on that thread was "Can I make it louder, or customize the sound? Now that would be something!" Other users commented on its conditioning effect:

> OK yes but in my defense I like watching my husband game so I hear that sound and it's time to hang out.
>
> Even my dog looks at me in disappointment when he hears that beep.
>
> My dog loves it! I can't mute it because how else will she know it's cuddling time?
>
> Mine does the exact same thing! He will run from anywhere in the house when he hears the beep and climb under the blanket to sleep between my legs. It's like our little routine.
>
> Yep! My dog recognizes the UI sounds of PS5 button + cycling over to "shut down" + putting the controller down a second later.
>
> Contrary to most people here I love the beep. Because when my dog hears the beep she knows what's up and lies down on my bed, waiting for me to join her.
>
> The beep doesn't bother me. It's the sound on the user select screen before it switches to headphone audio that gets me.

It's always worth finding out if your users want sound. As seen in the PlayStation example, some users like the boot-up sound—it's part of their experience and gears them (and their pets) up to play, offering a conditioning effect to not just the player, but others in their vicinity, signaling to all that a game session is starting. Other users don't want the sound at all and are angered by having sound forced on them. Most just want the opportunity to customize the sound or the option to turn the sound down or off. If you don't run user studies, how would you know how your users are responding to the product's sound? A simple user experience study would have informed the designers of the desire to include the option to switch it off or control the volume.

0.5 Who This Book Is For

A good user experience starts with knowing who your audience is. I hope anyone involved in launching a product today might benefit from this book, including the designers, the developers and engineers, and the UX research team. This book is written for the sound designers who want to understand how they might incorporate techniques drawn from user experience into their own process to provide clients with data-driven designs; it's also written for user experience researchers who want to understand how to incorporate sound into the user experience testing cycle; and it's written for the product designers or company owners who want to understand some basics of sound as part of their branding and user experience.

As described above, product sound design historically has often had the goal of reducing or eliminating noise that a product makes—for instance, making a vacuum cleaner motor sound quieter to meet standards and regulations or to make a more pleasing product. I am not covering the engineering of how to eliminate such noise here—for the most part, we're talking about adding, not subtracting sound. I am also not covering the design of audio products, such as how to improve sound quality in an MP3 player or set of headphones, or how to improve the sound of a violin, for instance. While these products make sound, as with noise reduction, there are many complicated engineering aspects that are simply beyond the scope of this book. This area of sound research is often referred to as *sound quality evaluation*. For further reading an excellent starting point is Nick Zacharov's *Sensory Evaluation of Sound* (2020). Finally, I'm also not providing a basic primer on user experience research. There are many books and resources available for those who want to learn general UX research practice (e.g., Gothelf, 2013; Marsh, 2022).

Rather than repeat material that has been covered in other books, my focus here is on UX research as specifically applied to sound-related questions. I am providing a blueprint for companies that want or need to design sound for their products but are unsure how to approach sound as part of their design process. I'm teaching you some testing methods to provide user experience data for stakeholders, as well as providing some science-backed data and talking points to help you to evangelize sound at your company. As noted, quite often sound gets ignored or relegated to last-minute decisions, and what results is often less than effective, ending up reinforcing the notion that sound is unimportant. This creates a vicious cycle, because it's nearly impossible to deliver high-quality meaningful experiences in such short turnaround times, so the sound never elevates the product to the level desired. Companies may then give up on sound before they have ever really invested in it to begin with, missing out on a great opportunity to increase their user experience.

As an academic working in sound for more than two decades, I'm especially interested in data-driven, science-backed research. I bring this background with me in my consulting work as I work with companies to incorporate and test sound in their products. As such, I am including hundreds of references and citations that the reader may want to explore in further depth. Sound is a relatively recent area of academic interest, and many questions remain unanswered, but the area is proving to be increasingly important in a highly distracted, multimedia world.

Having worked with executives at numerous companies, I recognize the need to educate, inspire, and provide concrete data about sound design. I believe this book will, therefore, provide some tools to the design teams who need to argue for a solid budget for sound, and demonstrate the process to go about exploring the best sound for their product. As I've seen through firsthand experience, product designers often don't even know what questions they should be asking about sound in their products, let alone the methods to go about answering these questions. It is my hope, therefore, that this book will provide some guidance on what types of questions to ask, how to answer them, and how to create the best sonic experiences for your products.

1 Talking Sound

Sound is vibration. Sound is therefore both something we hear and something we feel. Sound envelops us, and we can't shut it out—as Marshall McLuhan once said, "We simply are not equipped with earlids" (McLuhan and Fiore, 2003).

A few years ago, I spent a day without sound, wearing several layers of ear protection. We've all been blindfolded for fun as kids, or had to walk around somewhere very dark where we've temporarily lost our sense of vision, but what could be learned from a day without sound? I put in silicone earplugs, which promised a reduction of between 22 and 26 decibels (dB). I then wore some safety earmuffs overtop (the kind worn on gun ranges and construction sites), which also promised a reduction of 34 dB. With working inner ears, it's not possible to make yourself completely deaf—we hear through bone conduction even if we can completely block our ears—but this approach was pretty close. Unless there was a very loud sound, I was, for all intents and purposes, without sound for the day. I learned a lot about how reliant I was on sound unconsciously—even as someone who had spent my life studying sound. I recommend everyone try at least a few hours without sound during an ordinary day. It will make you realize how much sound is integrated into your daily life, without you ever having been aware of it.

1.1 Listening to Sound

There is a difference between hearing a sound and listening to a sound. We hear sounds all the time, but how often do we actually *listen* to sound? In the courses where I teach sound, the first thing I have my students do is to sit somewhere for five minutes and write down what they hear. A list might look like this:

- Fluorescent lighting hum
- Car driving past outside the window
- Someone walking down the hallway
- A scraping sound of someone shoveling snow outside
- The backup beepers on a truck at the construction site
- Whirr of the heating duct

There are a few common patterns that recur each year. First, most students miss out on the sounds that they themselves make—the writing of their pen on paper, their own breathing, or other sounds. The second is that everyone uses different language to describe sounds, and most students use multiple different approaches. Some sounds lend themselves well to onomatopoeia, for instance. Other times students may even try to draw a sound with a quick imagined sketch, or use a hand gesture to describe the way the sound moves through time. I try to get students to then go back and repeat the exercise, getting far more descriptive about the sounds the second time around. For instance, let's say in their first list they wrote, "a car driving past." The problem with such a description when used to describe sound is that it tells you only very roughly what sound is being heard. The sound is different depending on the type of vehicle being driven (a pickup truck sounds very different from a sports car), what the weather conditions are (tires in rain and snow sound different than tires on dry road), what time of day it is (a car in the middle of the night is going to appear to sound louder and potentially more reverberant, depending on the surroundings), what the speed of the car is, perhaps what gear it is in or if it is changing gears, what the mechanical condition of the car is, what kind of tires or exhaust it has, and other factors. Without all of this detail, we might conjure up a generic concept of *car-ness*, but it's not a very accurate descriptor of what has been heard. How the car is moving is perhaps an even more important indicator of what is happening: Are they squealing tires with some bass thumping out the windows, or are they creeping past very slowly, eerily, suggesting some form of surveillance? These are very different sounds and, played back, would give rise to quite different emotional responses. We have to agree on what the description of the car means to even begin to guess all of the associations with the sound of "a car driving by." Talking about sound is, therefore, a challenge, and I'll come back to this point several times.

How we talk and think about sound influences how we use sound in our creative processes. In *sound libraries*—collections of sound effects recorded for use by sound designers—sound effects are often categorized based on what caused the sound: "airplane" sounds, for instance, or "bird sounds." But sounds *could* be categorized based on what we might use them for: "scary sounds" (foxes, flies, or crow sounds are often used in horror movies), or "morning" sounds (the crow of a rooster or the dawn chorus of songbirds). When we design sounds, we often use sounds that are not tied to their actual source. For instance, we might record the snap of frozen celery sticks to create the sound of breaking a bone. Who would think to look for "vegetable sounds" in a sound effects library for their horror film unless they were aware of these uses? So, while causal descriptions can be useful to some extent, they are very limited when talking about sound's meaning.

1.2 Sound Attributes

As my students' responses suggest, we can listen and attend to sounds in different ways, and the language we use to talk about sound is muddy at best. Even after we've grasped the language, chances are we're going to have to talk about it with someone who hasn't yet learned that language. As children, we're taught a lot about visuals. We learn shape and color and texture, and the language to talk about these. If I asked you to draw a circle with a diameter of five centimeters and fill it in with a solid lime green color, you could probably come up with something very similar to what I have in my mind. What if, instead of the circle, I want you to copy the sound I'm imagining instead? I could try to make that sound myself, but the human vocal range is limited, and imitating sounds is a skill that takes time to develop. We run into one of the key difficulties in using sound as part of our design process: How do we describe and share what we imagine, when we don't have the words for such things? Sound has many words to refer to similar concepts, and very few that everyone agrees upon. Nevertheless, it's worth learning a few key concepts so you can discuss sound with your team, and try to understand the language that they speak.

There have been countless attempts to define a lexicon of some sort to describe sounds. We have so many different types of words that we use to describe sound that it is difficult to come to an agreement on how we use these words or what types of words to use. We often refer to our other senses to describe sounds—for instance, we'll call a sound *rough* or *thick* or *sharp*—and I'll discuss these *multimodal* interactions and how to leverage them in the next chapter.

Many people who have been trained in music will tend to use musical terminology and speak of specific pitches, chords, tempos, and other attributes using musical language. Of course, this language is limited also, since it was mostly designed for Western musical notation purposes, and is particularly poor when it comes to non-tonal sounds, the timbral properties of a sound (i.e., their texture), or recording and production techniques.

Using acoustic terminology, such as referring to frequencies and envelopes, can be the most objective and literal way to describe sound, but it doesn't tell us how a sound is being perceived by a listener. Sound can cut to the heart of our emotions, and there are many ways a 1,000 Hz sound can be perceived based on other attributes, context, and so on.

At a basic level, our sound processing goes through several stages, or filters (figure 1.1; see also Zacharov, 2020). First, a sound stimulus occurs: The sound can be measured using objective tools such as sound pressure level (SPL) meters or spectrograms. Next, the sound is received by our senses (our ears, brain, body); this requires no effort on our part. Our ears are all slightly different, and we have individual deficiencies and abilities, as well as experiences that make us more or less sensitive to the stimulus. Our auditory system is intrinsically linked to areas of our brain governing emotion and memory, such as the hippocampus and the amygdala. Indeed, different areas of our brain are even activated by unfamiliar versus familiar music (Pereira et al., 2011; Ara and Marco-Pallarés, 2021). We are also fooled into hearing certain aspects of a sound based on our other sensory inputs (see chapter 3, section 3.5, "Multimodality and Intersensory Integration"). This stage of measurement is also objective, although the results will depend on the individuals involved (see, e.g., Oxenham, 2018). The sound then enters our *sensory memory*. Sensory memory is short-term spontaneous memory, the first stage in our processing of information. When it comes to imagery, our visual sensory memory lasts about 0.1 to 0.2 seconds. However, with auditory sensory memory, estimates of duration are up to ten seconds

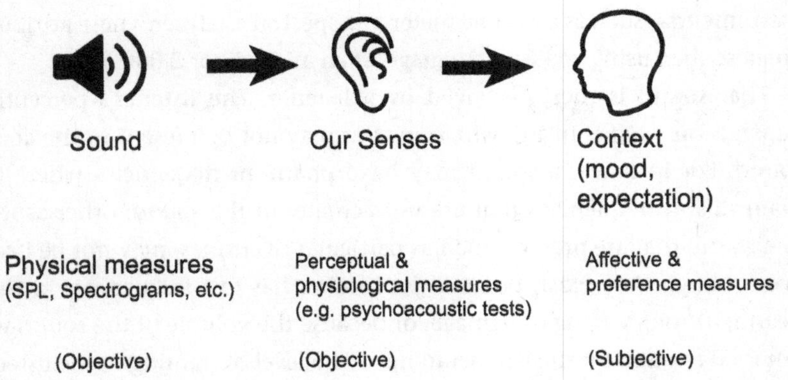

Figure 1.1
The stages of our response to sound.

long. During this brief period in sensory memory, the sound is processed at a basic level. This processing involves basic features of the sounds, such as pitch, volume, and timbre. The field of psychoacoustics is largely concerned with this stage.

Following sensory memory, our attentional processes determine which auditory information will be further processed. Our brain is alerted to particular sounds or particular aspects in the sound based on our background, the context, our expectations, our interests, our moods and emotions, or other factors. If attention is directed toward the sound (i.e., we're listening actively), it may move on to short-term memory for further processing. We determine if the sounds are encoded as speech or music, or if they have some other semantic meaning based on learned information. Once in short-term memory, sounds may undergo further processing, such as semantic encoding or integration with other sensory information. At this final stage, we may experience emotions, affective responses, subjective preferences, and mental imagery based on the sounds we have heard. Sounds are decoded, which can be influenced by many factors such as cultural influences, context and environment, and moods and emotions. These measurements are typically subjective, and will be individually differentiated.

We might, therefore, think about sound and its meaning as a series of events that occur where the language we use to describe the experience may change at any one stage of the process. We can begin with acoustic metrics, which are objective, scientific measures of the properties of the sound. It consists of several acoustic properties that can be measured using technical instruments—such as a decibel meter or a spectral analyzer. These attributes are described using acoustic language, such as 65 dB or 2,000 Hz.

That sound is then perceived by a listener. The listener's perception depends on what is heard, which may or may not be present in the actual sound. For instance, a sound may have phantom frequencies where the brain fills in frequencies that are not actually in the sound; other aspects of a sound that are present, such as particular overtones, may not be heard even though they exist, because the listener has lost frequencies in their hearing through aging or damage, or because the volume of the sound was not loud enough for the listener to hear. The level of training of the listener will play a role in what is perceived. What is perceived, in other words, has both elements of subjectivity and elements of objectivity. The descriptions here may include somewhat subjective measures such as *high frequency* or *loud* or *short* or more objective descriptions like *A4* (musical note) or *about one second*. Perceptual measurement is an attempt at an objective quantification of the sensory strength of individual sensory descriptors of a perceived sound.

From the perceived sound, we may have a conscious or unconscious emotional, affective, or cognitive evaluation of that sound. Sound can impact us even when we are asleep (Pedemonte et al., 2014). Sound evokes responses in a listener beyond the technical, acoustic, musical, or purely descriptive attributes. Here we may find affective descriptions like *rough, longing, harsh,* or some specific connotations (e.g., mental imagery) like *red sparkly bright shoes* or *teenagers chanting in a circle*. All the visual, tactile, and action-based imagery that the sounds evoke in the mind, along with the emotional and affective response that the sounds evoke in the listener play a role in their descriptions. These are entirely subjective responses, but there are usually many commonalities between listeners in these types of responses.

Finally, there may be specific meanings associated with those evocations in the context of the sound's use or reception. We may like or dislike the sound. We may find that the *harshness* we associated with the sound in the context of a song or a movie might become associated with *anger at the system* or the brightness of a sound might become associated with cleanliness and high-tech: In other words, it is a secondary level of signification that through its context may become altered with a personal interpretation. Since these interpretations depend on the listener's experiences and cultural backgrounds, there may be more of a differentiation between listeners' descriptions at this stage. They are entirely subjective, although there are often common references between people's responses.

It is not only the physical properties of a sound, in other words, that can determine our response to sound: The source of the sound and the relevance it has to the individual (along with many other factors, such as the mood they are in) can also influence perceptual response (Bergman et al., 2016).

It is difficult to quantify which elements of a sound may be influential in terms of their communicative properties, and thus designers can often only guess as to what properties may underlie particular user judgments about the product, meaning we need to run tests to explore the responses. User testing can therefore be a time-consuming, highly iterative process. The designers must understand how sounds may map to vague user judgments or connotations, and this can take some extensive testing to uncover.

There are a number of attributes, or parameters, of sound that each influence our response to that sound. A brief overview, along with some of the terminology commonly used, will be helpful to refer to with sound designers. An additional glossary is provided in the book.

1.2.1 Amplitude/Volume and Dynamics

Amplitude is an acoustics term for volume. But where volume tends to be measured in subjective ways (e.g., too loud!), amplitude tends to be measured in somewhat more objective ways (decibels). There is an element of subjectivity in the measurement of decibels (and thus there are many ways to weight the decibel scale), in part because our ears hear different frequencies at different volumes, but in general, we can use the common decibel measure, A-weighted decibels (dBA), which is usually just referred to as decibels (dB). There is a tendency among those not versed in sound design to always want sound to be louder, to have more impact through volume. But too much volume, especially over any length of time, can be off-putting or tiring to listeners, and amplitude is not always the answer to achieve an impact: A good sound designer has many other tools in their belt besides increasing amplitude. A clear example of this ability is the advertisements that sound louder than the television programs they accompany, even though they may be the same amplitude: In this case, they primarily use dynamic range compression to appear louder, without actually being louder (Moerman, 2004). Tolerance for volume can change depending on other properties of the sound and the context (Punch et al., 2004). Some studies have shown that among tones varying in loudness (40–90 dB), 50 dB was the most pleasant, but this would depend on where the sound was heard and what background ambient levels were. Moreover, it also depends on the frequency of the sound: The study found that lower frequencies were heard as more pleasant until about 60 dB, where higher frequencies were heard as more pleasant (Todd, 2001). Speaking very broadly, happiness is associated with medium-high sound levels, and small variations in sound levels. Quiet environments are considered more pleasing than loud environments (although context is important) (Cunningham et al., 2023). Sadness is associated with a constant low-level sound volume. Anger is associated with high sound level and some loudness variability. Fear is associated with low levels, large variabilities, and rapid changes in volume. However, these cannot be assumed to be rules on their own, since there are so many variables at play.

1.2.2 Envelopes and Duration

Sounds exist in the temporal domain—in other words, they change over time. How a sound changes over time is described using what we call an ADSR envelope, or just *envelope*. ADSR stands for attack, decay, sustain, and release, and it describes how a sound's amplitude begins (the attack), and how it changes over time (the decay and sustain), before describing how it ends (the release). For instance, a hand clap has a very fast attack, a very short decay and sustain, and a short release. However, the envelope will change based on the environment we're in: If we go into a large, cavernous space, the attack of our hand clap may be the same, but the sound will linger through reverberation, so the sound will have a differently shaped envelope, with a long decay, sustain, and release (figures 1.2 and 1.3). A bowed musical instrument

Figure 1.2
An approximate envelope of a short, sharp sound like a hand clap in an average room.

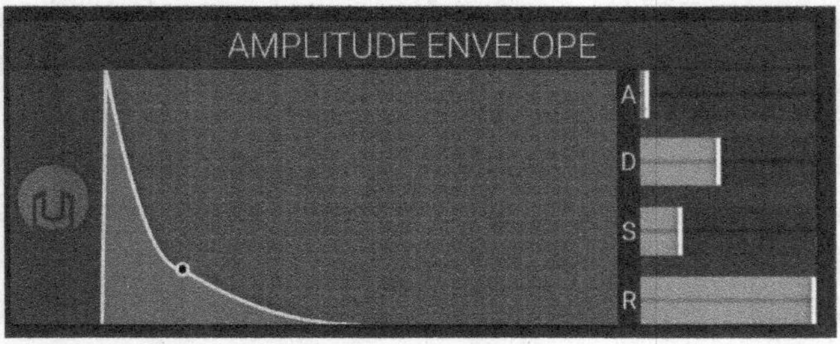

Figure 1.3
An approximate envelope of the same clap sound in a more reverberant space.

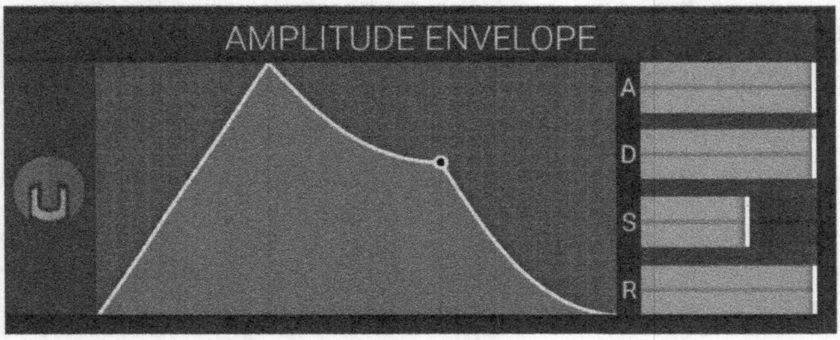

Figure 1.4
An approximate envelope of a violin being bowed. Note the slower attack and drawn-out decay.

like a violin, on the other hand, has a very long, drawn-out attack (usually), longer decay and sustain, and longer release (figure 1.4). Recent research has shown that the envelope of a product sound—particularly its release time—can influence emotional ratings (see Bhatara et al., 2025).

1.2.3 Pitch and Frequency

Frequency is used to describe how *high* or *low* a sound is: It is the acoustic corollary to pitch in music (note, too, the metaphorical nature of describing a sound as high or low). Frequency is a more technical term than pitch, though, and we don't use musical notes like A or C; we use exact measurements in hertz (Hz). Humans have a frequency range of about 20 Hz to 20 kHz. In musical terms, we can also talk about the octave, or the register in which a band of frequencies occur. For example, a piano can be divided into several registers, such as the bass register (lower frequencies), mid-range register (middle frequencies), and treble register (higher frequencies), or into octaves (with 88 keys, a standard piano has a little over 7 octaves). Each of the registers has a distinct tonal quality—think of the icy tinkling of the very high keys, for instance. Similarly, human voices are categorized into different registers, such as tenor, bass, and falsetto, each producing different frequencies and tonal qualities.

The difficulty with frequency as a term, though, is that most sounds we hear (in fact, all except artificially generated sounds) are not a single frequency. Sounds in the natural world consist of not just one frequency, but a series of harmonics or overtones—other frequencies that are often related in a mathematical way to the dominant frequency, and are quieter than the dominant one, but which give added texture, or grain, to the sound. With some sounds it's quite easy to identify a dominant frequency, but with other sounds—what are sometimes referred to as *noise* sounds—there is no dominant frequency or, at least, not one we can easily discern.

Frequency has been suggested to correlate to size, in what is referred to as the *frequency code* hypothesis (Ohala, 1983). A study found that both male and female professionals spoke in deeper and more resonant voices when answering questions that required specialist knowledge, a behavior likely intended to project their authority. Notably, women lowered the pitch of their voices to a greater extent than men when giving career advice rather than directions, possibly because women may need to lower their voices more to have an effect on their listeners, or because men typically speak almost as low as they are able. Additionally, recordings of career advice in deeper voices were rated as both more authoritative and more competent than recordings of voices in a higher frequency in a follow-up study, even though the listeners did not understand the content of what was being said (Sorokowski et al., 2019). In this way, Margaret Thatcher was taught to lower her voice, to appear more authoritative. Her biographer Charles Moore wrote: "Soon the hectoring tones of the housewife gave way to softer notes and a smoothness that seldom cracked except under extreme provocation on the floor of the House of Commons" (Gardner, 2014). Such sound bias is not limited to human communication: Studies of primates and other species have noted that animals elicit shrill and staccato sounds during aggressive moments, and legato, smooth sounds during positive social interactions (Rendall et al., 2009).

Studies have shown that tones of between 200 and 1,000 Hz, which correspond roughly to the human speaking voice frequencies, are perceived as the most pleasant (Todd, 2001). Higher frequency sounds are often associated with lightness, smallness, and sharpness. Sharpness, typically in the 2–4 kHz range, is generally associated with unpleasant and annoying sounds (Cunningham et al., 2023). These higher frequency sounds can also evoke feelings of excitement, energy, and even playfulness. High-frequency sounds might be linked to concepts like youth, speed, and delicacy (I say *might* because there is more than just frequency involved in the connotations). Sounds with lower frequencies are often associated with heaviness, largeness, and depth. They can evoke feelings of seriousness, stability, and darkness. Low-frequency sounds might be linked to concepts like strength, depth, and power.

1.2.4 Instrumentation, Articulation, and Orchestration

Instrumentation refers to the use of different instruments to create musical sounds. The instrumentation may be *guitar and bass*, for example, and most genres are distinguished at least in part by their instrumentation choices. Orchestration refers to assigning specific musical parts to individual instruments within an ensemble. The way instruments are played—their articulation—can also have a significant influence on feel. Pizzicato (plucked) strings on stringed instruments can introduce a whimsical or light, playful feel. Brass fanfares give a sense of importance and majesty. Arpeggios can create a sense of nostalgia or wonder. An ostinato (a riff) can give a sense of drive, momentum, or energy.

1.2.5 Timbre

Instrumentation choice will often define the timbre of a band's sound. Timbre is kind of like the texture of the sound. Sometimes timbre is referred to as the quality of the sound, or the tone of the sound, particularly when it comes to musical sound. Timbre includes elements such as how many and what ratio of frequencies (overtones, harmonics) there are in the sound, as well as the envelope (see above), which can change how that texture is perceived. Other elements in the sound, such as vibrato (a subtle back-and-forth wobble in the frequency, known as frequency modulation) or tremolo (a trembling back and forth wobble in volume, known as amplitude modulation) can also contribute to timbre.

1.2.6 Rhythm and Temporality

Rhythm is the basis of all music, and refers to the pattern of durations, accents, and time intervals within a musical composition. It is the element of music that gives it a sense of pulse and movement. Rhythm is created by the arrangement of different note values, pauses (rests), and accents, which together form a structured sequence of events within a musical piece. We can also discuss rhythm in terms of the *rhythmic density* of a piece: the complexity of the rhythm and how many different rhythmic elements are used.

Tempo, in music, refers to the speed or pace at which a musical piece is performed. It is measured in beats per minute and determines the rate at which the underlying pulse or beat of the music occurs. Tempo has a significant connection to the human body, as it often mirrors and influences our physical and emotional responses.

Tempo is often not just the main *beats* in the bass and/or drums of the music, but also the *surface rate,* the rate of occurrence of notes that may be on top of the beats. As musicologist Philip Tagg (n.d.) puts it, "Put crudely, tempo is the 'boom-boom' factor and surface rate the 'diddly-diddly' factor."

The tempo of music often aligns with the natural rhythms of human movement (Dahl, 2005). For example, a fast tempo might encourage and reflect energetic and rapid movements, while a slow tempo might lead to more deliberate and relaxed motions. Fast tempo is associated with high arousal and tension, surprise, happiness, potency, and anger, and a decrease in heart rate variability (van der Zwaag et al., 2011). Speaking very broadly, slower tempo is associated with sadness or serenity.

1.2.7 Tonality and Noise

The *tonality* of a sound refers to how musical a sound is perceived, which means having one dominant musical note or pitch. A noise, for instance, is typically a sound that has no discernible dominant pitch, and no tonality. In music, tonality refers to the organized system of pitch relationships within a piece of music that centers around a specific pitch called the tonic. It is a fundamental concept in Western music that provides a sense of stability and orientation within a composition. Tonality forms the basis for melodies, harmonies, and musical progressions, creating a hierarchy of pitches that gives structure and coherence to music. Tonality is essential for creating melodies and harmonies that are aesthetically pleasing to the ear.

It establishes a framework for musical expression, allowing composers and performers to convey emotion, tension, and release through the manipulation of tonal relationships. Tonality heavily influences mood and emotion.

1.2.8 Melodic Contour/Directionality

Melodic contour refers to the shape or trajectory of a melody's pitch movement over time. It describes how the musical notes in a melody rise, fall, or remain constant in pitch, creating a distinct pattern that contributes to the melody's overall character and emotional expression. Music generally has a sense of directionality—it appears to go somewhere (typically toward a cadence), and has a beginning, middle, and end, and therefore creates a sense of anticipation. It tends to tell a story, in a sense, and creates a series of expectations through tension and release.

The cadence is where the melody rests or concludes, creating a sense of closure. Cadences contribute to the overall flow and structure of the melody. Disruption to that goal can lead to a sense of frustration and lack of resolution. Slot machines, for instance, often lead players up in a melodic contour where you feel like the next spin might give you a resolution, but if it doesn't, it encourages you subtly to keep playing until you do get a resolution (a win), and hear the close of the cadence (Dixon et al., 2015). Melodic contour and directionality can typically be drawn out on paper, and can also often be felt as a sense of gesture, with roundness or sharpness, ups and downs (Tagg, 2004). A rising pitch might be associated with ascent, growth, or positivity, and is associated in product sounds with a confirmation (Bhatara et al., 2025). It can evoke a sense of progress, optimism, and movement. A falling pitch might be associated with descent, decrease, or negativity. It can evoke a sense of decline, melancholy, or completion. Our tendency to rise in intonation with a question, and to use falling or flat pitch contours for statements, is a universal tendency across languages (Bolinger, 1978).

1.2.9 Verticality and Harmony

Harmony and verticality refer to layers of notes in a sound or piece of music, and therefore include the arrangement of notes in vertical space and over time. The harmony includes all the notes that work with the melody. For instance, a melody may be the main hummable theme of a piece of music, but the bass in that music may play a counterpoint to that melody. Sudden changes in harmony are often used to grab a listener's attention. In most popular songs, for instance, a change in harmony is used to signal the chorus of a song and differentiate it from the verse. In Western music, harmony is usually in a key, or potentially a mode. Minor keys are associated with lower arousal and sadness, but this is not universal at all (van der Zwaag et al., 2011). For instance, in music not part of the Western classical music tradition, like folk music, minor keys can be happy (think of sea shanties like "What Shall We Do with a Drunken Sailor," which are often in a minor key). Thin, simple harmonies are associated with high valence (i.e., pleasantness).

1.2.10 Spatiality

There are lots of contributing factors that help a sound or piece of music create a sense of three-dimensionality, which is one of the most important ways sounds differ from visuals. Unlike visuals, which occur only in front of our eyes, sound can permeate around us, creating a sense of envelopment and immersion.

We can discuss spatialization in terms of perceived locality (width, depth, height), of separation between the perceived sources of the sounds, of definition (by creating a sense of space between sounds so that they can be individually differentiated), or of envelopment (typically, created through bass, since bass is difficult for humans to localize, so creates a feeling of being enveloped by or immersed in the sound). We can also discuss spatiality in terms of reverberation and perceived reverberance (i.e., how long the sound persists after the source has stopped) through environmental reflections. This can be measured on a scale of dry (no to little reverberance) to wet (highly reverberant), or in terms of the time it takes to decay. The perceived three-dimensional location in space of a sound is created through a variety of techniques (see, e.g., Collins, 2015).

We might also add here the terms *stereo*, *mono*, and *binaural*, and *surround* sound, which describe not the *perceived* location, but the *actual* location of the playback through loudspeakers or headphones. The physical location can reinforce or contradict the perceived location, as in phantom imagery (see, e.g., Nishiguchi et al., 2020), the precedence effect, or the use of images to confuse location, as in the ventriloquist effect (e.g., Vroomen et al., 2001).

1.2.11 Mixing and Production

Mixing is the process of combining individual sound elements of a composition into a final, cohesive output. The goal of mixing is to establish a balanced, dynamic, and sonically pleasing creation that captures the intended artistic and emotional qualities. Mixing is an artistic and technical process that requires a trained ear, a deep understanding of sound, and proficiency in audio engineering software and equipment. A well-executed mix enhances the overall impact of the sound/music, allowing the listener to differentiate each element clearly and experience the intended emotional depth and sonic qualities of the composition. Mixing is sometimes synonymous with equalization (EQ), or balance, since adjusting these properties is perhaps the most important role of the mixer.

1.2.12 Effects

There are a variety of effects that are either incorporated into a recording or, more commonly today, added to the mix in post-production. It's unusual for music today to be released without any effects: Even before analog effects processors, artists were recording in reverberant spaces (e.g., Elvis even recorded in the bathroom—see Doyle, 2004). Some specific effects include the following.

Reverb: Describes the effect of simulating the acoustic environment or space, adding reflections and ambience to the sound. Makes the sound feel like it's in a big room or space. Can create a feeling of warmth and envelopment, or atmosphere, depth, and spaciousness. Some terms used to describe reverb include spacious, echoey, roomy, immersive, cavernous.

Delay: Refers to the effect of creating an echo or repeating sound by delaying the original signal. Can create a sense of distance and dimension. Some terms include echoing, repeating, bouncing.

Phasing: Describes the effect of altering the phase relationship between two similar audio signals, creating a sweeping or swirling sound. Creates an otherworldly, surreal, or dreamy feeling. Terms include swirling, spacey, dreamy, otherworldly, wobbly, psychedelic, druggy.

Flanging: Similar to phasing, flanging refers to a type of modulation effect that creates a distinctive jet plane or swooshing sound. Terms include jet-like, airy, wavy, swooshy, fluid.

Chorus: Makes the sound thicker and more spacious by adding multiple layers of the same sound, often with an effect or slight delay between the tracks. Can create a sense of richness, depth, and lushness. Terms include hick, lush, harmonious, ensemble-like, shimmering.

Distortion: Refers to the effect of intentionally altering the sound's waveform, adding grit or saturation. Can add a feeling of aggression, edge, intensity, grit, and/or energy. Terms include gritty, crunchy, grungy, edgy, raucous, dirty, rough, fuzzy, noisy.

Filter: Describes the effect of modifying the frequency content of the sound, often used to emphasize or attenuate specific frequencies. Adjusts the sound's tone, making it brighter or darker. Can be described as warm, mellow, bright, dark, subdued.

EQ (equalization): Refers to the process of adjusting the balance of frequencies in a sound to enhance or shape its tonal qualities. EQ fine-tunes the sound by adjusting bass, treble, and other levels. Terms include balanced, clear, crisp, boomy, tinny.

Compression: Makes the loud parts softer and the soft parts louder, reducing the overall dynamic range. May be described as evened-out, controlled, punchy, tight, consistent.

Bitcrushed: Refers to an effect that intentionally reduces the bit depth of a sound, creating a lo-fi or gritty quality. Gives a retro sound, like old video games. Terms include lo-fi, retro, chopped, 8-bit, old-school.

Panning: Describes the process of placing a sound in the stereo field, moving it from left to right or vice versa. Terms include moving, panoramic, wide, dynamic, spatial, ping-pong.

Tremolo: Refers to a modulation effect that rapidly varies the volume of a sound, creating a pulsating or vibrating effect. Can add movement and rhythmic interest. Terms include pulsating, vibrating, shivering, rhythmic, quivering.

Vibrato: A frequency modulation musical technique used primarily by vocalists and instrumentalists to add expression and depth to a sustained note. It involves a regular, pulsating variation of pitch and intensity, created by oscillating the pitch slightly above and below the target note. Vibrato can add emotion, warmth, and a sense of life to a musical performance. Happiness is associated with a medium-fast vibrato, sadness with slow vibrato. Terms include whimpering, tremulous, quavering, wavering, quivering, undulating, fluttering.

Wah-wah: Describes an effect produced by a filter that sweeps the frequency range, often used with guitars. Terms include expressive, vocal-like, funky, wailing, playful.

Pitch shift: Refers to the effect of altering the pitch of a sound, creating higher or lower notes. This can create a feeling of otherworldliness, depending on how it is done. May be called altered, warped, shifting, chipmunk voice, bending, pitch-bent.

Reverse: Describes the effect of playing a sound backward, producing a unique and ethereal texture. Can create a surreal, mysterious sound. Descriptive terms include backward, reversed, retraced, rewound, mysterious.

Gating: Refers to the process of cutting off the sound below a certain threshold, creating a staccato or rhythmic effect. Can be used to create emphasis. Terms include choppy, staccato, rhythmic, stuttering, gated.

Modulation: Refers to the application of effects that involve changing the parameters of the sound over time, such as vibrato, chorus, or flanging. Ring modulation refers to a metallic, robotic-sounding effect.

1.2.13 Fidelity and Artifacts

Fidelity refers to the quality of a reproduction in relation to an original, and can also include the overall clarity. Low fidelity introduces distortion, crackle, muddiness, and other artifacts into a sound. Some aspects of fidelity can bring to mind a sense of nostalgia. To create the warm analog feeling of vinyl, we can reduce the higher end of the frequencies, and add some crackle, for example.

1.3 Conclusions and Take-Aways

As can be seen from the long list of terms we use to talk about sound production and reproduction, sound analysis and design require some considerable expertise. There is a lot to learn, and it can intimidate people who are not familiar with this language. However, it can be helpful if everyone on the design team has a shared vocabulary to discuss sound. The language we use to talk about sound can influence our creative practice, as shown. We can get highly technical, or very metaphoric and creative in the ways we describe sound, and each method has its own advantages and disadvantages. Technical descriptions can help convey precise information about the characteristics of sound. This can be useful where specific measurements and parameters are important. Technical descriptions can ensure that everyone involved in the design process understands the sound requirements and objectives clearly.

Metaphoric descriptions can help evoke emotions, associations, and sensory experiences related to sound. This can be beneficial in user experience design, where the goal is to create products or environments that engage users on an emotional level. Metaphoric descriptions can help designers communicate the desired emotional impact of the sound and align their vision with the overall design goals. Using other modalities (such as touch, taste, smell, or vision) as metaphors to develop comparisons in other senses can be very helpful, since we often integrate sounds with other senses, as will be discussed in chapter 3. A user may not understand that a sound is high frequency, but may call it *sharp* or *piercing*. Translating the metaphors used by people not in the sound industry can be an important part of the interpretation of user information, but the metaphors themselves can tell us a lot about how a sound is being perceived.

2 Sound and Design

This chapter provides an overview of product sound design, starting with a brief history. The idea of using sound to solve design problems has been around for thousands of years, and a brief understanding of the ways in which sound has been used is useful to begin thinking about why, when, where, and how to use sound today. I then cover the types of product sounds that can be designed (beyond the mechanical sounds that require engineers), to get you thinking more about the ways you can leverage sound in your products.

2.1 A Brief History of Product Sound Design

Recent years have seen tremendous growth in companies that are now paying more attention to the sound design of their brands and products. This awareness of sound has grown as part of a shift in focus to a product's overall user experience, and of the multiple senses that contribute to that experience, in what is commonly called *sensory design*.

It is likely that one of the earliest forms of designed sound beyond musical instruments was some sort of alarm system to alert us to the presence of others. We use doorbells and burglar alarms for this purpose today, but there is recorded evidence of people employing animals such as geese and dogs as early warning systems going back to at least 386 BCE, as recorded by Roman historian Livy, who described using birds to alert the Romans to the approaching Gauls (McCrie, 1988). In ancient Greece, the water clock, known as a clepsydra, had mechanical whistling birds at specific intervals to mark the passing of time, an early example of integrating sound with a utilitarian function. Other commonly designed sounds include the use of bells to signify various events: In medieval England, a curfew bell was rung to signal that it was time for everyone to cover their fires and go to bed. Musical bells had been used in villages in Europe since about the year 400 CE and would signify various important events, for instance, a call to prayer, a death, or a wedding. Bells were not just signifiers of events, however, as they were also symbols of power (Gutton, 2000). In ancient China, the number of bells you had was directly tied to your rank. The emperor could have bells on all four sides of his palace, and a duke or prince would have three, whereas an official could have just one (and the rest of us peasants wouldn't get any!), clearly indicating the associations between noise-making and power (see also Schafer, 1994).

By the industrial revolution of the late 1800s, sounds could be found in a wider array of products. The talking doll was invented by Thomas Edison in 1877, who reportedly later admitted that "the voices of the little monsters were exceedingly unpleasant to hear" (Söderqvist and Bencard, 2010). Later, toys like the 1928 Phil the Radio Dog robotic electric dog could bark. The telephone, invented in 1875, is still a ubiquitous sound today, of course. But sounds were also added to cash registers as alarms by the 1890s, and coin-operated vending machines began to use sounds and even voices by the 1920s (Marcosson, 1945). In the late 1920s vending machine boom, a cigarette machine by the Consolidated Automatic Merchandising Corporation would deliver a package of cigarettes with a "Thank you" along with the brand's slogan. Chocolate manufacturer Stollwerck's clucking hen vending machine distributed chocolate in the 1920s along with a clucking sound. But sound effects were much more commonly used in other forms of amusements like mechanical midway/arcade games: A bell was introduced to pinball games in 1933, with Harry Williams's Contact game. It was reported that "he set up one Contact with a bell in a playing area next to one without. Then he sat back to see what would happen. The machine with the bell collected twice as many nickels and Harry Williams knew he had a new feature" (see Collins, 2016).

By the 1930s, as industrialization and urbanization increased, public awareness of noise grew, and there were many political and public actions against sound, including the formation of the Anti-Noise League in the United Kingdom in 1933: "Motor cars and their hooters, pneumatic street drills, clattering lorries, banging doors, aeroplane engines, and the new and growing plague of wireless loud speakers—there is no end to the volume," as noted in the *British Medical Journal* (Willson, 1933). New York mayor Fiorello LaGuardia declared a war on noise in 1935. That same year, the Science Museum in London held a temporary exhibition on noise abatement, after lobbying by the Anti-Noise League. Exhibits expounded the futuristic approach to products like elevators or typewriters that had been re-designed as quiet or silent (Mansell, 2017).

The attempted elimination of noise was a more common practice than adding noise to products for several decades afterward, although this removal of sound can create its own problems: "People want quiet appliances, but at the same time they equate noise with power. How do you make a quiet food processor that sounds powerful enough to handle the job?" (Wolkomir, 1996). At times, attempts to remove sound have backfired, as sound was seen as integral to the product's optimal functioning. In the 1940s, Remington introduced a silent typewriter, but nobody wanted to buy it. Author Darren Wershler-Henry (in Hansen, 2007) tells us:

> This kind of new technology that Remington assumed would be a real boon to typewriter users and ended up not being the case at all. The keys produced this kind of unsatisfying dull thud, not at all that sort of clatter of keys that we hear every time we hear a news bulletin and about to start. By the point that that typewriter was manufactured, fleet reporters were so used to the sound of keys hitting the platen of their machine that they wouldn't type if they couldn't hear them. And pretty much all writers who have talked about the typewriter, up to and including Henry James, have said, is that they need that kind of audio feedback of listening to the machine in order to actually compose successfully. . . . He actually couldn't write until he heard the typewriter operating beforehand. It was basically—it had become the trigger that enabled him to writer instead of a kind of side effect of the writing itself.

Sonic interfaces were built into computers as early as the late 1940s (for a history, see Shintaro, 2012). These would soon both feed into and take inspiration from science fiction: The beeping sine waves of interfaces in 1950s movies were drawn from the electronic creation of sine wave and white noise sounds in old computers (see *The Magnetic Monster* (1953) or *Gog* (1954)). The development of the ceramic piezoelectric buzzer to generate sound in the 1950s and 1960s saw the technology work its way (with its ubiquitous beep) into many products in the 1970s and 1980s, becoming an iconic sound associated with cheap products today (see Gopinath 2021).

As technology has moved to more digital production, we have kept some of the mechanical sounds of earlier devices as skeuomorphic design—the concept where digital design mimics the real-world equivalents, such as keyboard sounds or the click-whirr of a camera taking a photograph on a phone. With the expansion of digital sound and affordable speakers, we have also been able to add sounds where we couldn't before, and design whatever sounds we want, rather than be limited to electronic beeps, and this is where product sound design has really taken off. Today, our appliances sing a song to tell us the washer or dryer cycle is finished, the microwave or rice cooker is done cooking, the door of the fridge has been left open, or the brewing is done in a coffee maker.

2.2 Types of Product Sounds

Product sounds encompass various types of sounds that provide feedback, communicate information, and enhance the overall user experience. There are probably more opportunities to design sound for your product than you anticipated, including not just the user interface, but the name of the product or brand, branding sounds (including sonic logos), hold music, interface sounds, packaging sounds, and ambient and environmental sounds.

2.2.1 Phonosemantics: The Name of the Product or Brand

Phonosemantics is the study of the relationship between the phonetic features of speech sounds and their associated meanings or emotions (the sound symbolism). Phonosemantics suggests that certain sound patterns in language can evoke particular sensory, emotional, or conceptual impressions. However, while there are undoubtedly instances of sound symbolism in various languages, phonosemantics remains a topic of debate in linguistics. Critics argue that the relationship between sound and meaning is often arbitrary, and heavily influenced by cultural and linguistic factors. Furthermore, context plays a significant role in shaping the perception of meaning, and phonosemantic associations can vary widely between languages and dialects. Nevertheless, if you begin to make a list of pharmaceutical drugs, you'll soon see some patterns in their names: Drug companies are heavily invested in the idea that certain names evoke certain imagery.

Dan Jurafsky, in *The Language of Food: A Linguist Reads the Menu* (2014), describes how the sound of a name influences our purchasing decisions. For instance, vowels are divided between front vowels and back vowels. Front vowels include i (as in mach*i*ne), and ɪ (as in sl*i*m) with the tongue high in the front part of the mouth. Back vowels are produced with the tongue lower, in the back of the mouth, such as ɑ (as in l*a*rge), o (as in c*o*ld) and ɔ (as in m*o*re). Front vowels are often (but not always) associated with smaller, thinner, lighter things, and back vowels with larger, fat, or heavy things.

While phonosemantics remains a somewhat controversial area, it is important to consider the name of your brand or product in terms of the types of imagery or associations that are evoked by its phonemes. Names have certainly been shown to influence attitudes toward the brand (e.g., Keller, 1998).

2.2.2 Sonic Branding and Sonic Logos

Sonic branding aims to evoke specific emotions, enhance brand recognition, and reinforce the brand's values and attributes through the use of sound. Sonic branding can create a consistent and cohesive brand experience across touchpoints, convey information, and evoke moods quickly. When it comes to the brand, all aspects of product sound should be considered as potential sources for sonic branding, including those mentioned in this chapter, but also in terms of advertising (e.g., television, radio, internet), associated mobile apps, events (held online or in person), and other opportunities.

How many of us, as children, took pleasure in re-creating the well-known logos to which we were constantly exposed? Looping our arms around our heads in a circle and roaring to the MGM logo, or barking along with the TV sign-off, "Sit Ubu Sit. Good dog. Woof!" of Ubu Productions. These popular sonic logos are so well recognized that they are often parodied in television and film: The MGM roar has been spoofed countless times, from *Monty Python* (1969) to *The Muppets* (1981). The Ubu logo has been satirized on *Robot Chicken* (2006) and many other shows.

Sonic logos—also called *mnemonics*—need not be a melody, specifically, but rather may also contain a unique timbre/texture, rhythm, or other sonic attribute associated with a brand. The THX logo, for instance, is highly identifiable (parodied on *The Simpsons* (2007) and *Bluey* (2020), among others) but has no melody as such, and neither does the Netflix "Ta-dum" (see Khosla, 2020).

While sonic logos are not new (MGM began using its lion roar in 1924), their recognition as an important part of a brand's overall strategy has been relatively recent. Mercedes, for instance, first introduced the visual Mercedes Benz logo in 1909, but it was not until 2007 that they first designed a sonic logo. Today, sonic logos are still relatively uncommon compared with their visual counterparts, but are increasingly being used as part of an overall audio branding strategy, as an important component of a multisensory approach to brands.

When it comes to some brands, audiences are not only *aware* of the sonic logos (as indicated by the parodies discussed above), but actively engaging with them. There are countless collections on YouTube of logos for video game companies, computers, and other beloved products. These videos get thousands (or even millions) of views: People are actively searching out advertising to watch, listen, and collect, something most advertisers can only dream of. Some companies have even attempted to bring attention to sonic logos by involving customers in their design (such as Sennheiser's 2008 sonic logo competition), to limited success.

Perhaps one of the most famous sonic logos today is that of Intel, whose "Intel Inside" mnemonic forms what musicologist Philip Tagg terms a *transscansion* (Tagg, 2005). A transscansion is a series of musical notes or sound effects that stand in for the prosody of speech of syllables in a word or phrase in such a way that the repetition of these sounds may offer a substitution for the actual word(s) in our minds. You are likely familiar with the bouncing ball singalong, a technique of old Max Fleischer cartoons using an animated ball that keeps the beat with the lyrics. A transscansion works in a similar way, whereby the rhythm implies the syllables or words often visually shown on screen.

The Intel sonic logo's composer, Walter Werzowa, describes the process of creating the sound by repeating the words Intel Inside and singing them: "'Intel Inside' is four syllables, so four notes. And it's the fourth and the fifth [musical intervals] that are the most common intervals in every culture. I put them together with a little divider note—a 'clink'—at the beginning and played it for [friend Kyle Cooper]. We presented it to Intel and they loved the idea" (cited in Steiner, 2025). In his video analysis of "Intel Inside" that includes a voice-over to make the transscansion explicit, Tagg notes, "Hearing and seeing the synchronicity of the spoken words 'Intel inside' added to the existing synchronicity of the visual logo and the four-note jingle creates, I would argue, an awareness of the mnemonic mechanism implicit (and intended) in those three seconds, an awareness that goes way beyond the mere intellectual knowledge that such transscansions with such functions do exist. The added dimension of my repeating the spoken words 'Intel inside' is intended to make explicit the implicit character and automatic mnemonic mechanisms of that particular transscansion" (Tagg, 2008). Indeed, I would argue that it is now difficult to think or say the words "Intel inside" without calling up the melody in one's mind.

Transscansions are also found in other sonic logos: In the United States, NBC television has a sequence of three chime-like musical notes as if to reinforce the letters NBC in our minds. James Bull (2009) discusses his creation of the Vodafone transscansion, which consisted of, initially, three notes (Vo-da-fone), but found that when asked to sing the notes back, listeners could only remember two. When a fourth note was introduced at the start as a kind of trigger, the listeners would then remember the three notes. In other words, listeners needed an indicator trigger sound that would draw their attention, before the transscansion—such a trigger is also present in the Intel Inside logo, which actually consists of five sounds.

There have been many studies to substantiate the idea that recognition and recall of a brand are improved by associating text (a slogan or the brand name) with a melody (see Yalch, 1991). Research has also found evidence that music facilitates the recall of a phrase or slogan, even when nonsense syllables are used (Serafine et al., 1984). Of course, not just any music can be added to a slogan or phrase—it has to be congruent with the message (North et al., 2004).

Joel Beckerman, founder of sonic branding agency Man Made Music, admits that "many sonic logos, especially in the form of snippets of sound dropped in almost exclusively at the ends of ads, are meaningless. . . . They tell you nothing because you feel nothing. Without an associated long-form anthem, a sonic logo has no emotional memory to trigger" (Beckerman, 2014, p. 65). Minsky and Fahey (2017) provide the claim (apparently from their own market research) on the recognizability of sonic logos, such as their French Railway SNCF logo. They assert that 92 percent of listeners correctly identified the brand by sound, and 88 percent on hearing just two of the notes. Graakjaer and Bonde (2018) also determined that one or two notes in a distinctive musical style (i.e., with a unique timbre) were sufficient enough to gain recognition and target a specific social or demographic group, creating a "whole nexus of social and cultural values" with a product.

The research into the effectiveness of sonic logos is varied in hypotheses, approach, and method. Sonic logos are usually experienced as part of an audiovisual message, and a number of studies have attempted to explore the relationship between sound and visuals. Anzenbacher (2013) suggests that if visual and auditory information were presented in a consistent, simultaneous way, there would be a "considerable increase of neuronal activity," indicating that the combination of a visual image with a sonic logo should be more effective than sonic logo or image alone. This suggestion is in keeping with other research into the use of melodies more generally in advertising and learning described above.

Current research into the effectiveness of sonic logos is somewhat conflicted. The research appears to show that sonic logos increase brand recall and likability, but only if congruent with brand message. One of the difficulties with existing research is teasing out what, exactly, makes a sonic logo effective, how often it should be used, or where/how it should be used. There are some general trends in the research that point to possible characteristics of an effective sonic logo, including the following:

- To draw attention, frequencies should be around 1 kHz, with a short attack.
- The ideal number of tones in a melody is around six.
- An upward melodic contour is preferable.
- Pitch is more important than rhythm to recognition, but timbre is the most decisive element.
- High-energy sounds like percussive tones (e.g., marimba) help to draw attention.

2.2.3 Start-Up Sounds

Start-up—or boot—sounds are, of course, the first sounds heard when a device is powered on. One study into start-up sounds in computers and vacuum cleaners found that the sound's impression impacted the repurchase intention (Kato et al., 2020). Having just purchased two new laptops (a generic PC and a new MacBook), I must say there was a distinct pleasure in opening the Mac for the first time and hearing the Mac boot sound as I raised the screen. The PC experience was simply not comparable, not least because it involved many more steps before hearing the start-up sound. The Mac sound is, of course, iconic now: Pixar's *Wall-E* (2008) even included it to mark the robot as a Mac product. The absence of the sound in some Apple products meant, as its sound designer Jim Reekes said, "now that there's no start-up sound, it's like sitting down at a restaurant and there's no one there to greet you."[1]

2.2.4 Interface Sounds

The concept of interface sounds has evolved hand in hand with the advancement of technology. Interface sounds can help to foster a more intuitive, engaging, and accessible interface. Many interface sounds have taken inspiration from science fiction on-screen, such as *Star Trek* (1966–) or *Minority Report* (2002) (see Shedroff and Noessel, 2012). The beeping and whirring of starship controls and the electronic blips of futuristic computers have become iconic and set the standard for what audiences expect from technology. As interface sound designer Peter Drescher put it, "it is now standard procedure in Hollywood movies that computers and mobile devices *must* produce interesting, responsive, futuristic, interface sounds, even when it makes no sense for them to do so" (Drescher, 2011). Movies set expectations for your users, so if interfaces are noisy on-screen, then they are expected to be noisy in real life.

The 1980s saw the introduction of interface sounds into our home computer software, with Bill Gaver's SonicFinder application (1989), designed to add sound to the graphical user interface (GUI) of the late 1980s Apple Macintosh. Today, interface sounds are expected, even when many users may turn them off. Jordan Novet, in an article for *CNBC*, describes how smartphones helped to drive the increase in sounds in interfaces, as tech companies began hiring sound designers (Novet, 2021). Even though interface sounds may be switched off, or users think they don't want sound, they still serve important purposes in some circumstances. According to Conor O'Sullivan, Google's lead sound designer, the biggest challenge for a sound designer is to "play sound for people that don't necessarily always want to hear sound or don't want at least to notice these sounds all the time."[2]

Sounds are important to some functions of interfaces: For instance, my colleagues and I explored the role of health-related sounds in interfaces in games, and found that the sound cues were important to relate health-related information to players, despite players having an on-screen visual cue (Robb et al., 2017). In this case, where the eyes have many places to look and the user is visually distracted, sound can augment the experience. Interface sounds have been shown to influence the perception of *social surrogacy* (the imagined presence of other people around), and have been shown to increase willingness to pay, higher adoption, and likelihood to recommend (Poupis et al., 2021).

In the research, interface sounds are often divided into auditory icons and earcons (Brewster 1992): *Auditory icons* are sounds that are designed to represent objects, functions, or events. Auditory icons exhibit a clear and intentional resemblance to the real-world sounds or concepts they represent, such as the use of the camera click sound on mobile phones. This resemblance can be achieved through pitch, rhythm, timbre, and other auditory parameters. Similarly to how visual icons in user interfaces are graphic symbols that quickly convey meaning, auditory icons are sonic symbols that convey meaning through their resemblance to the real-world sounds they represent. They have the benefit of being easily recognizable, allowing users to quickly associate the sound with its corresponding meaning or action without effort.

Earcons, on the other hand, consist of sequences of musical notes or patterns used to convey specific information or actions. Earcons are typically composed to be distinctive and easily recognizable, allowing users to develop a learned association of the sound with specific meanings or actions. Earcons are useful to build up a hierarchical or branded family of sounds that can be learned. When speech is added to earcons, these are known as *spearcons*. A meta-analysis found that spearcons were superior to auditory icons or earcons when it came to reaction time and accuracy (and auditory icons were superior to earcons). When it comes to subjective ratings, earcons also rank the worst in terms of preference in research studies (Nees and Liebman, 2023). However, these studies typically have not accounted for the conditioning effect that longer-term exposure leads to. Moreover, the studies have rarely employed professionally designed sounds to test, and therefore the results should be taken with a grain of salt.

2.2.5 Hold Music

Hold music is a tricky area, and most of us have learned to hate hold music, a response to the fact that we are often already frustrated by the time we come to hear it. It is, almost inevitably, associated with a negative experience. Nevertheless, music makes the experience of being on hold appear to make time pass more quickly (Adkins, 2021). Hold music is an opportunity to brand this experience, but on the other hand, since it's associated with frustration, you may not want to brand this particular time with your users. It depends, of course, on how long your users are going to be on hold for and for what purpose. People are probably more likely to associate positive feelings with being on hold for three or four minutes to sign up for some new product, but are far more likely to experience negative feelings if they've been left on hold for twenty minutes to fix a product issue or to cancel a contract. Moreover, the quality of the music is often very poor, and put in short loops, making the experience even more painful than it needs to be. There's nothing worse than hearing Pachelbel's Canon in twenty-second loops with interruptions of "Your call is important to us" for an hour. Music does decrease the perceived amount of time spent on hold, but different types of music impact the rate of hanging up (see Ramos, 1993; Guéguen and Jacob, 2002). An article in *Wired* magazine states that Apple has cultivated somewhat of a cult following for its choice of hold music, with one ex-employee even creating a Spotify playlist of songs chosen.[3] Legend has it that Apple CEO Tim Cook listened to the music on Apple's line after a customer complaint, and changed the entire experience. I recently experienced an optional hold menu with Bell Canada. Unfortunately, there was no option for no hold music, only for which genre I wanted. And then, the song ran out and looped hundreds of times before I finally got through—not a great sound experience.

2.2.6 Packaging Sounds

First impressions matter. An often-overlooked opportunity for sound design lies in the packaging of products. In recent years, unboxing videos have become very popular on YouTube. These are videos in which a fully packaged product is slowly unveiled, revealing not just the product, but the packaging that it was sold in. These videos perhaps began with electronic products like iPhones, but today unboxing videos can be found in every category of product. Most of these videos are created by customers, but brands themselves have started to showcase product unboxing. The primary driver of unboxing videos, according to Google (Think with Google, n.d.), is that they create anticipation and excitement; however, some consumers just want to see what is inside the packaging so that they know what they are going to buy, especially when shopping online.

Researchers have shown that packaging can influence a user's perception of (and potential intimidation by) an unfamiliar product, such as a robot, and packaging shapes the subsequent interactions with a product, even influencing task performance (Cafaro et al., 2016; Schlüter et al., 2021). This effect is a demonstration of what is known as the *halo effect*—the idea that an impression in one domain will influence the impression in other domains. In other words, a brand that has invested in high quality packaging design will have their product perceived to be of a higher quality. Packaging design is of growing importance as manufacturers recognize the value of creating an initial emotional response, to the point where some companies (like Apple) even own patents on their packaging (Craig and Cunningham, 2017).

In *Fast Company* magazine, Elvis Hsiao writes, "The box, like the product it once held, stands as a testament to the design values Apple espouses, a physical embodiment of a philosophy that prizes simplicity, functionality, and beauty" (Hsiao, 2023). Some brands are also taking advantage of the ASMR trend, knowing that some sounds are highly pleasing to some people. ASMR, or autonomous sensory meridian response, is the tingles or chills that many people get from a sense of intimacy, which can often be invoked by using sounds recorded with a very close microphone. The One Co in Cheshire, UK, for instance, posted their new packaging with an ASMR video on their Facebook page, writing, "Sound on!! What better way to start showcasing our brand new packaging then [*sic*] with some ASMR. We can't wait to show you the rest!"[4]

Research backs up the importance of packaging sound: The sound of a cork coming out of a wine bottle was found to impact taste and quality ratings of wine (Spence and Wang, 2017). Manipulating the sound of an aerosol spray can made a product appear more pleasant (Spence and Zampini, 2007). There is even evidence that the sound your spoon makes when it rubs against the side of a product's container has an influence: One study found that yogurt is perceived as sweeter and more expensive when it is contained in glass jars rather than plastic or clay (Lerma et al., 2017). Potato chip manufacturers put their chips in intentionally noisy packaging, and the chips themselves have been the study of many research projects, which have found that amplifying the crunching sound of the chips increased the perception of the chips as crisp. Sun Chips famously had packaging that was so loud that it prompted complaints by consumers, leading to them change the packaging (Gura, 2010).

Premium packaging can drive repeat purchases, and packaging has become a part of the user experience (Hsaio, 2023). Attractive packaging activates the reward-seeking and impulsive areas in the brain (Hubert et al., 2013). Velasco and Spence (2019) point out how there are different choices made when products are sold with different packaging, and the difference in sound can influence the consumer's decision at the store to purchase a particular brand. They describe how the sound of Snapple bottles opening became part of Snapple's branding and created a sense of anticipation in purchasers. But it is not just in food products, and not just at the initial point of sale, that sounds matter. In an article published in the *Wall Street Journal*, a packaging designer at skincare and makeup manufacturer Clinique describes the many prototypes that go into creating a mascara wand whose *click* emits a sense of luxury (Byron, 2012). One author described how designers at Hermès spent "three seasons finding the perfect sound a bag should make when it shuts" (Bowles, 2013). Packaging design expert Evelio Mattos (2022) explains that "if you're trying to communicate quality . . . or trust . . . sound [in packaging] plays a bigger role than you think."

2.2.7 Interaction Sounds

The way your product interacts sonically with the environment should also be taken into consideration. For instance, in one study the type of sole used in high-heeled shoes and the type of flooring (ceramic vs. carpet) in the store influenced the personality traits associated with the shoe wearer, for instance, as more or less honest, kind, or intelligent (Klanovicz et al., 2022). In other words, it may be worthwhile to investigate the potential to manipulate your product materials in order to alter the product's sound interactions with the environment, or at least to test your product in the various environments in which they are used, especially if you have dedicated stores. How does it sound on a wood table versus Formica or granite? How does it sound after it gets wet? How does it sound outside versus inside? Depending on your product, these interactions may make for a positive or negative experience, and small details can make a big difference: Adding rubber or felt pads to the bottom of a laptop, for instance, may make your laptop computer feel more solid and be quieter to type on.

2.2.8 Environment Sounds: Servicescapes/Atmospherics

Many of us hear sound and music as background, or aural wallpaper, in a variety of environments: Indeed, it would be strange, if not disconcerting, to go to a restaurant or store today and not hear some music being played. In recent years, the importance of designing the whole experience of an environment—whether a retail environment, a restaurant, a service environment, or other environment in which the customer has some form of interaction—has become a part of sensory marketing. Some stores have their own signature scents, and the music or sounds played (at least, in many chains or larger brands) is carefully chosen. The *servicescape* offers an important opportunity for sonic branding, for emotional engineering, and even for changing customer behavior.

There have been many studies of the impact of sound and music on different environments. Choice of music influences patronage and store choice (familiarity and/or liking of music has a positive effect, for example), increases impulse buying, and alters the speed at which customers will consume food or make purchases (slower tempo and lower volume will lead to longer stays, and increased the speed of bets in a casino) (Garlin and Owen, 2006; Trompeta et al., 2022). A meta-analysis of tourism and hospitality servicescapes revealed that the presence of music had a small but significant effect on customer reactions and responses, with a significant positive effect on setting evaluation. Another meta-analysis found that the presence of music increased the pleasure felt, that higher volumes led to greater satisfaction (up to a point!), that consistency between music and brand led to increased positive perceptions, and that the presence of music led to higher spending and enjoyment (Biswas et al., 2018). The tempo of the music had some influence: Slower music led to perceptually shorter wait times (in queues, for example), and custom music led to greater perceptions of brand fit and more favorable overall impressions, increasing time spent in the store (Falkenberg et al., 2021). Studies have shown that the congruence between the environment and expectations is particularly important in these cases.

In addition to trying to promote positive experiences, sound and music can also be used to reduce negative experiences. Stores have used music to deter teenagers or delinquent behavior: 7-Eleven stores in British Columbia in the 1980s began playing classical music or Muzak at stores in the evenings to deter loitering (Azpiri, 2022). Abercrombie & Fitch play music between 85 and 90 dB to keep older customers away (levels this high have also been shown to impede cognitive processing, perhaps leading to more impulse decisions). Moderate levels in the range of 70 dB, however, increase likelihood to purchase innovative products when compared with lower levels of sound (Mehta et al., 2012). Witchel et al. (2013) decreased antisocial behavior in Brighton by altering the soundscape of a whole street in the entertainment district.

The servicescape experience is translated into the online world as well, with virtual or e-servicescapes. I suspect as we move to virtual or augmented reality-based web shopping experiences, the role of sound will become much more important in the virtual world to fill in the gap left by the loss of touch and smell. Background music and ambient sound effects will influence how users react with a virtual space, as my former grad student's prior work has demonstrated in video games: We conducted a study that found the background music in a game had a variety of impacts on the players' emotional states, and influenced their actions. The players consciously or subconsciously attempted to make connections between the music and the game's narrative, events, imagery, and playing tactics. Players found coincidences between elements of the music and actions on-screen. When music was incongruent, players reported less ability to play, less ability to concentrate, and less immersion in the game (Wharton and Collins, 2011).

It's not only music that should be a consideration in environments, but also sound effects. Francis et al. (2017) found natural sounds like bird songs and wind influenced listener's attention and sense of well-being. On the other end of the spectrum, low frequency background sounds altered customers' comfort levels, priming a threat response and increasing anxiety, leading to risk-avoidant consumer choices (Lowe et al., 2019). The physical environment of the store itself can also play a role—the type of flooring can change the bodily sensations of shoppers, and the sound of the footsteps can influence behavior (Bresin et al., 2010).

2.3 Conclusions and Take-Aways

Historically, designed sounds served practical functions, such as alarms and time markers, evolving through industrialization to modern digital skeuomorphic sounds in various appliances. In recent years, companies have increasingly focused on sound design as part of sensory design to enhance overall user experience.

Sonic branding, through a variety of different avenues, including the product and brand name, aims to evoke emotions, enhance brand recognition, and reinforce brand values through sound, creating a cohesive brand experience across various touchpoints, including advertising, apps, and events. Effective sonic logos, or *mnemonics*, increase brand recall and likability when congruent with the brand message, often incorporating unique timbres, and rhythms as transscansions to stand in for syllables in a phrase. Research highlights the importance of attention-drawing frequencies, a limited number of tones, upward melodic contours, and distinctive timbres for recognition.

Iconic start-up sounds, like the Mac boot sound, enhance the user experience and brand recognition, creating a distinct pleasure compared with generic devices, and their absence can feel like a lack of a welcoming gesture.

Interface sounds enhance user experience by making interfaces more intuitive, engaging, and accessible, drawing inspiration from science-fiction cinema to meet user expectations for responsive, futuristic sounds. While users may sometimes turn them off, these sounds play crucial roles in functions like health-related cues in games, social surrogacy, and user engagement, though their effectiveness varies with design quality and user conditioning.

Sonic experiences beyond the immediate product can also influence the reception and perception of the product. Effective hold music can potentially enhance the user experience if it is of high quality and appropriately selected. However, branding hold music is risky, as it can reinforce negative associations, especially during long waits for resolving issues. Packaging sound design is crucial for creating positive first impressions and can influence a consumer's perception of product quality. Packaging sounds are becoming part of the user experience, enhancing the unboxing process and shaping subsequent interactions with the product.

The sound a product makes when interacting with different environments can influence consumer perceptions and personality traits associated with the product. It is beneficial to test a product in various environments to understand how it sounds in different settings. Small modifications, such as adding rubber or felt pads to a laptop, can significantly improve the user experience by making the product quieter and more stable. These details can enhance the overall perception of product quality and usability.

In modern environments such as restaurants and stores, background music and sounds are integral to the customer experience. This practice, known as sensory marketing, enhances the overall experience and can influence customer behavior. Familiar or liked music increases patronage and impulse buying, and can alter the speed at which customers consume food or make purchases. The tempo and volume of music also play significant roles: Slower music can make wait times feel shorter, while higher volumes can increase satisfaction and spending, up to a point.

The principles of sound design in physical spaces also apply to virtual environments. Background music and sound effects in virtual or augmented reality settings can significantly impact user emotions, actions, and overall experience. The congruence between sound and the environment is crucial for positive customer experiences. Incongruent sounds can lead to decreased immersion and negative perceptions.

Understanding sound's role in different contexts can help in designing more effective and engaging customer experiences. Testing and optimizing these sonic elements can enhance user experience and drive positive brand associations. And while careful sound design is the hallmark of a great brand, knowing when *not* to use sound is as critical as knowing when to use it. Silence can connote luxury (think of the inside of a luxury vehicle and how quiet it is compared with many cars), but as shown, when introduced where people want feedback, silence can be the wrong choice. The best way to know what your users want is to test different versions of your product and see what works best for them.

3 The Psychology of Sound

We rarely direct our attention to sounds in our everyday environment, which means that sound often works on a subconscious level. Yet even without our conscious awareness, sound can impact our behavior, our emotions and moods, our experiences, and even our health (Moesgaard et al., 2020). Our brains often determine if sounds are meaningful without our being aware of the process—for instance, our brains learn to tune out recurrent sounds in our environment like clocks or watches ticking, the hum of electricity, or cars driving by (this is referred to as *auditory adaptation*) (see Willmore and King, 2023). Our blood pressure and stress levels can increase even when we hear sounds while asleep (Lechat et al., 2022)!

There is more research into sound and its impacts than I can cover in a single chapter; however, I have summarized a few key areas of particular relevance to product sound design. It should be noted that I am presenting the latest information available, but much of the research into our brains, neuroscience, and the cognition and psychological aspects of auditory processing is very much a work in progress. Scientists are constantly updating or developing theories based on new work.

3.1 Sound and Behavior

Sound can have a significant influence on human behavior, and one of the ways it does so is through the concept of classical conditioning, often referred to as Pavlovian conditioning. The effect is named after Ivan Pavlov's experiment with dogs, where he conditioned them to salivate at the sound of a bell. Even in the absence of food, due to the repeated pairing of the bell with food delivery, the ringing sound of the bell would cause the dog to respond. This phenomenon demonstrates how specific sounds can elicit automatic and learned responses, shaping behaviors in conscious, subconscious, and unconscious ways. Just like Pavlov's dogs, we have also become conditioned to respond to sounds in certain ways: the doorbell or our phone ring, for instance, may trigger anxiety or happiness.

Operant conditioning takes things a step further: Where Pavlovian conditioning pairs two stimuli to create an association (e.g., the bell and the food), operant conditioning relies on consequences (reinforcement or punishment) for actions or behaviors. In operant conditioning, positive reinforcement involves presenting a pleasant stimulus after a behavior to increase the likelihood of that behavior being repeated (such as with a reward sound in a game). Negative reinforcement involves removing an unpleasant stimulus after a behavior to increase the likelihood of that behavior being repeated (such as removing the beeping of a seat belt sound in a car once you buckle up). There are also two types of punishment: positive punishment and negative punishment. Despite their names, both forms aim to reduce undesirable behavior, but they do so in different ways. Positive punishment involves presenting an unpleasant stimulus after a behavior to decrease the likelihood of that behavior happening again. Negative punishment involves removing a pleasant stimulus after a behavior to decrease the likelihood of that behavior happening again.

Media like games and gamified apps use sounds as positive or negative reinforcers in a form of operant conditioning. For example, sound often serves as immediate feedback for players' actions, indicating success like a *level-up* (a positive reinforcement), or failure, such as the beeping of low health in Nintendo's *Legend of Zelda* games (a negative reinforcement). There is also positive and negative punishment—in stealth games like *Metal Gear Solid*, an alert sounds when the player is spotted by enemies, discouraging being seen by encouraging stealthy behavior (a positive punishment), and in *Animal Crossing*, when the player hits a villager with a net, the usual pleasant music stops (a negative punishment). Such feedback serves to reinforce the player's actions, making it more likely they will stop or repeat the behavior that led to that outcome.

In summary:

- Positive reinforcement is adding a pleasant sound to increase a behavior, such as a pleasant tone or musical sequence when a dishwasher cleaning cycle finishes, encouraging users to unload it.

- Negative reinforcement is removing an unpleasant sound to increase a behavior, such as a continuous beeping sound that stops when the fridge door is closed properly.
- Positive punishment is adding an unpleasant sound to decrease a behavior, such as a harsh beep when the kettle is turned on without water, discouraging users from running it dry.
- Negative punishment is removing a pleasant sound to decrease a behavior. For instance, if I had an electric toothbrush that has a timer, I might disable the pleasant end-of-brushing tune that sounds if the user does not brush for the recommended two minutes, encouraging them to brush for the full duration.

Reinforcement cues can be very powerful. My prior work with the gambling research team at the University of Waterloo explored how sound on slot machines impacted player physiological responses, and their subsequent behavior. There are no negative sounds on most slot machines—the player only hears positive feedback on wins, and silence on losses. We discovered that the machines were deceitful, however, in that they also contained *losses disguised as wins*. These were times when the player won back less than they bet, so technically it was a loss, but the machine treated it as a win by playing the winning sounds. For instance, if I bet a dollar but won back fifty cents, I've lost fifty cents overall, but the machine treats it as a win. Players responded physiologically to the losses disguised as wins in the same way that they did with wins; that is, the sonic cues signified winning and induced the same response as wins. Players were therefore hearing a lot of winning sounds, and no losing sounds, which led them to overestimate how many times they'd won versus lost. The players were conditioned to respond to the sounds through their association with winning, even when they heard the sound but were not playing the game, reinforcing the betting behavior (Dixon et al., 2010).

Sound effects in movies and games can evoke emotional responses and influence behavior. A sudden loud sound in a horror movie, for example, can trigger a startle response and make the audience more engaged in the suspenseful storyline. *Jaws* (1977) is a great example of conditioning in movies: very early on we learn to associate the sound with the shark. Later, the expectation that the association creates is disrupted, as we both get the sound without the shark, and at other times the shark without the sound, adding to the fear created. Indeed, sounds paired with aversive stimuli can very quickly induce fear responses (LeDoux, 2000). The association often takes only one or two times to be reinforced enough for us to respond to the sound. Such films also make use of what is called the *acoustic startle reflex*, with sound-based jump scares when very quiet scenes are paired with very sudden, loud scenes (Davis et al., 1997).

Beyond physiological responses, sound can also influence more subtle and complex behaviors. For instance, as described in the previous chapter, background music in retail environments has been strategically used to shape customer behavior. Upbeat music might encourage a faster shopping pace and a more positive mood, potentially leading to increased spending. Conversely, slower-paced music might encourage shoppers to linger and explore, fostering a more leisurely experience (Yalch and Spangenberg, 2000). Even the mere presence or absence of music can influence us. Pleasurable sound, such as music we enjoy, is tied to the *dopaminergic* system in the brain—the reward system that encourages us to seek out pleasure (e.g., Madison and Schiölde, 2017). Recent neuroimaging studies have found similarities between how music influences the brain and how our brain responds to other rewards: In simple terms, music can activate the same reward center in the brain as addictive substances (Mas-Herrero et al., 2021).

3.2 Mental Imagery

Many of us conjure images in our minds—that is, we create mental imagery—when hearing sounds or music (figure 3.1), although this is not universal: Studies suggest just under 10 percent of people have never experienced visual imagery to music (Day et al., 2020). Such imagined experiences can be so vivid that they closely emulate the experience of actual perception in our brains, and can influence emotion and behavior (Kosslyn et al., 2001). Mental imagery is perhaps the key mechanism behind the role of emotional responses to, and engagement with, sound (Presicce and Bailes, 2019). Neuroimaging research has shown strong links between emotion and the production of mental imagery (Taruffi and Küssner, 2019).

Mental imagery is not just *visual* imagery, but can involve other senses (Küssner et al., 2023). Musical *earworms* are one type of musical imagery, for example (Jakubowski et al., 2017). In prior studies it's been found that not just music, but also sound effects can elicit images of the events or actions

Figure 3.1
Sound-evoked mental imagery: For most of us, we hear a sound, and our brain conjures up imagery, often of the causal action behind the sound.

behind the sound (Collins and Johnston, 2023). In studies of radio and mental imagery, the more sound effects a presentation contained, the more mental imagery was evoked (Soto-Sanfiel et al., 2023). High mental imagery radio advertisements were also found to increase audience involvement and attention, along with boosts in recall (Bolls and Lang, 2003). Studies have also found that audio-invoked mental imagery increases recognition for product-related information (Bolls, 2006). The theory suggests, therefore, that high-imagery advertising will be stored in our memory in a different way than low-imagery-evoking advertising, and that sound effects increase the level of imagery processing in ads, potentially leading to a stronger emotional reaction and increased messaging (Miller and Marks, 1992).

Mental imagery activates brain regions related to memory (Wu et al., 2006). This process can be trained, and expertise in an area leads to greater brain activity. For instance, film cinematographers and sound designers were studied using functional magnetic resonance imaging (fMRI) and asked to imagine sounds or images. The auditory cortex was activated for sound designers, and the occipital-temporal and parietal cortex were activated for cinematographers, suggesting a plasticity of the brain that adapts to what we are spending our time doing. The researchers could even decode the film's genre from brain activity (de Borst et al., 2016)!

Mental imagery can be conscious or unconscious, just as our everyday perception can be conscious or unconscious. What's even more fascinating, mental imagery changes the way that we perceive reality. When attempting to replicate some well-known cross-modal interaction illusions like the ventriloquist effect, the McGurk effect, or the illusory bounce percept (see section 3.5, "Multimodality and Intersensory Integration," below), mental imagery has also been shown to impact our perception in the same way, showing that our imagination can change our perception of reality (Berger et al., 2013).

3.3 Memory: Recall, Recognition, and Repetition

Sound has the profound ability to evoke emotions and memories. Hearing a song, many people can recall the last time that they heard that song, and what was happening, even twenty or more years after they have last heard it. People with a relative with dementia or Alzheimer's disease may have noticed the power that music has to aid memory recall (Vanstone et al., 2009). Even when other forms of memory are compromised, the emotional salience of familiar music can endure. Music can transport listeners to specific moments or phases of life, thanks to the connectivity between the auditory cortex, hippocampus, and emotion-processing regions of our brain.

Memory can also influence how we perceive and interpret a sound or a song. Our past experiences, emotions, and associations with specific sounds can shape our perception of them. For example, if a certain melody was playing during a joyful event in your life, hearing that melody again might evoke positive emotions and memories associated with that event. In this way, a sound associated with a pleasant experience from the past can change our mood when heard again in the present. On the other hand, if a particular sound was present during a distressing situation, encountering that sound in the future might trigger feelings of unease or anxiety (Juslin and Västfjäll, 2008). There are even websites that have gathered songs that have been *ruined* for people by gruesome movie scenes where the song appeared (e.g., McCann, 2019).

When discussing memory, it is worth differentiating recall versus recognition. In a *recall* task, people must reproduce what they have learned, searching their memory, as in an exam. In a *recognition* task, people are asked if they recognize the stimulus, such as if they've heard a sound before, comparing it internally with prior coded information. Both memory tasks can be useful in exploring user experience (Özcan and van Egmond, 2010). When visual information is presented at the same time as sound, as with a logo, for instance, recall is improved (this is known as the dual-coding effect). However, visual information can also reduce performance in recognition tasks, possibly since the visual information takes much of our attention and therefore we do not listen as well to the auditory information (known as visual over-shadowing) (Özcan and van Egmond, 2010). Our recognition of sounds can also be impacted by context: We will recognize the sound of a shaver in the context of a bathroom scene, for instance, but given the sound alone and out of context, it will be more difficult to recognize (Özcan and van Egmond, 2005).

Studies have shown that information presented audibly can be easier to remember than information presented visually, as our brains are equipped to learn sounds quickly to develop language skills. This phenomenon, known as the *phonological loop*, suggests that repeating information verbally or hearing it spoken aloud can improve memory retention (Derks-Dijkman et al., 2023). Melodies in particular are often remembered more easily than words or images. Jingles and even non-musical brand sounds can be recalled for years (Yalch, 1991): If I mention Oscar Meier wieners to many North Americans, they might immediately recall the jingle and sing it in their heads (or out loud!).

Our preference for (or liking of) music typically initially increases with exposure, in what is called the *mere exposure effect* (Zajonc, 1968). Familiar sounds can grab our attention, engage us in a product, and increase message processing and memory for products and brands (Raja et al., 2020). Generally speaking, we prefer familiarity over unfamiliarity, particularly when it comes to situations where we may be anxious (such as when making a major purchase, for example, or taking a test). A recent study found that familiarity was the biggest predictor of liking among music, regardless of the complexity of the music (Madison and Schiölde, 2017). Likewise, when it comes to products, people tend to choose a recognized product/brand over an unrecognized one (this is, after all, one reason why brands pay so much for advertising!). One study found that pairing brands with recognized music increased the likelihood of those products being chosen by consumers (Anglada-Torta et al., 2022). However, in tests of pairing products with music, music that was recognized but *disliked* meant the consumer would not choose the brand (Anglada-Torta et al., 2022). This connection can be complex. Lantos and Craton (2012) outline four interconnected factors in terms of advertising music: the music (e.g., genre, fit with brand, mood, familiarity), the listener (e.g., preferences, personality, culture), the situation (e.g., time of day, social context), and the advertising processing strategy of the listener (e.g., attitude toward advertisements, attention). To maximize the impact of music on consumer behavior, we must consider the impact of these on our target audience.

Although liking for a stimulus increases with initial exposure, it also decreases with overexposure. We may love a song the first few times we hear it, but after it's been played over and over, we begin to dislike it. Studies have found that liking increases in a linear fashion until about thirty-two exposures, where the U-shaped inversion occurs (Hunter and Schellenberg 2011). This inverted-U shape, known as the Wundt curve (figure 3.2; see Madison and Schiölde 2017), does not occur when music is just in the background, however; it occurs only when users are asked to focus on the music (Schellenberg et al., 2008). Being asked to pay attention to the music reduces the pleasure in far less time (Schellenberg et al., 2008).

The Wundt curve is also influenced by the complexity of the music, as well as personality type, in particular, openness to experience (Hunter and Schellenberg, 2011). For some, familiar music might be more enjoyable because it's predictable, while others might prefer more complex music because it's engaging and offers something new with each listen.

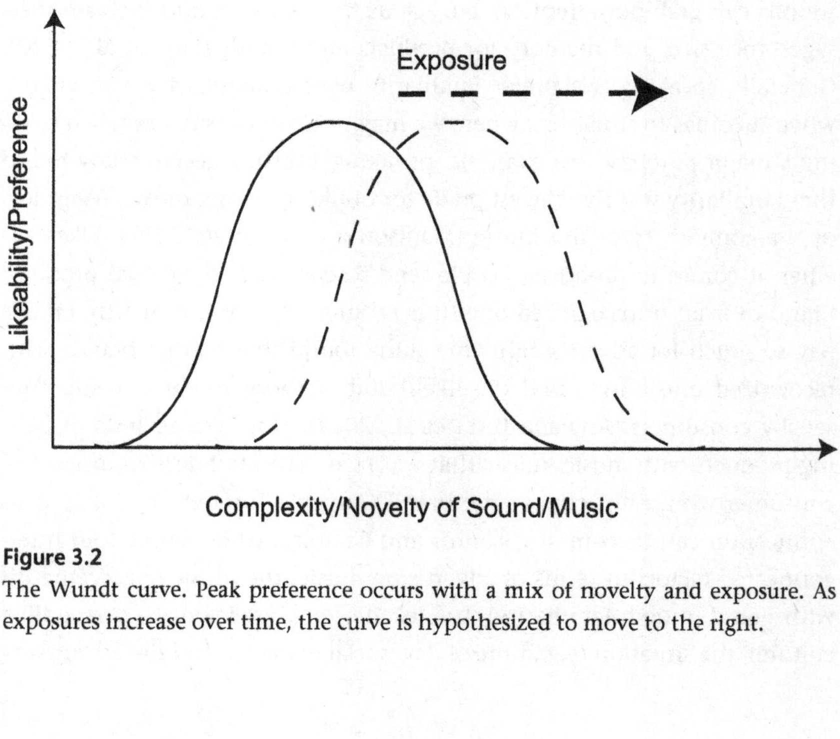

Figure 3.2
The Wundt curve. Peak preference occurs with a mix of novelty and exposure. As exposures increase over time, the curve is hypothesized to move to the right.

Repetition of some types of sounds can lead to *acoustic adaptation*, where the brain's response to that sound diminishes over time (Ulanovsky et al., 2003). This is often why a repetitive noise (like a ticking clock) might be annoying at first but eventually fades into the background (Grill-Spector et al., 2006). When a sound is consistent and non-threatening (like the hum of an air conditioner), our brains adapt and reduce the neural response to it, allowing us to focus on more novel sounds or tasks (Rabinowitz et al., 2011). But when a deviation from the predicted sound occurs, our attention is quickly drawn back (Fritz et al., 2010). Not all repetitive sounds are easy to tune out, of course: The seat belt warning beep, for instance, would be difficult to ignore, and has been designed to be that way by using a frequency our brains find hard to ignore.

3.4 Cognitive Load and Attention

Cognitive load refers to the mental effort required to process and manage information in a person's working memory. The cognitive load theory of attention suggests that the type of information provided, the type of task, task difficulty, complexity, and the individual's preferences can determine the mental effort required (Lavie and De Fockert, 2005). For example, to return to the example of driving, if we are driving in an unfamiliar city trying to find a street (without a GPS), we have a tremendous amount of stimuli—we have to pay attention to our driving, to the road conditions and other vehicles, cyclists, and pedestrians on the road, and simultaneously try to read the street name signs. Visually, we are taking in so much information that most of us find it difficult to also, for instance, carry on a conversation or listen to music. The additional auditory stimuli, in this case, is too much stimulation and is redundant or irrelevant information—it doesn't add anything to our needs at that time. On the other hand, with a GPS speaking aloud our upcoming turns and beeping if we exceed the speed limits, we can off-load some of the visual work (we no longer have to look at a map or our speedometer), reducing our overall cognitive load by redistributing some of our cognitive resources to another part of our brain.

Attention and cognitive load have been studied separately for many decades, but recent research is attempting to understand their impact on each other. Cognitive load is related to attention—we can selectively ignore some perceptual inputs in order to focus on others. In terms of competing sounds, we call this the *cocktail party effect*—we can focus on a conversation with someone in front of us, while there are multiple conversations happening around us that we can tune out. If we had to pay attention to multiple simultaneous conversations, we would have a very high cognitive load. Attention impacts our perception and response to sound (Steffens et al., 2020). One study compared a distracted group with a group asked to focus on environmental sounds. The distracted group rated the sounds as more pleasant than the group that was asked to focus on the sounds (Steffens et al., 2020). This result makes sense, since a sound that is interfering with a task is more likely to be seen as an annoyance.

Whether cognitive load is increased or decreased by sound depends on several factors, including the age of the listener (children are more likely to be distracted by new task-irrelevant sounds; Wetzel et al., 2021), attention to the sound, the complexity of the sound, what the sound is communicating, and if that sound is relevant (congruent) or irrelevant (incongruent) to other information. More complex tasks require more attention and thus are more subject to cognitive load difficulties (Caviola et al., 2021). Task-irrelevant sound usually increases cognitive load (Sörqvist et al., 2012). Complex, unfamiliar music or sound can compete with the attention required for a primary task, and therefore might overload cognitive resources (Juslin and Västfjäll, 2008).[1] Some recent research has shown, however, that irrelevant sounds that are emotionally distracting can increase alertness and improve performance (Max et al., 2015).

In studies into distraction and learning, where information that is provided in sound conflicted or distracted from the visuals, the result is usually a decreased capacity for learning (Moreno and Mayer, 2000). In such instances, usually one modality (typically vision) will dominate and the other(s) may be ignored and overridden by visuals, known as the *Colavita effect* (Colavita, 1974). In general, when provided with incongruent information between modalities, visual information will dominate most situations, although recent research has shown that this is not always the case (Ngo et al., 2011).

On the other hand, the use of sound increases learning in tasks where there is some partial overlap between cues—that is, where each modality (sound, visuals) provides some of the same information, but not all of the information (Noetel et al., 2022). Redundant information, that is, information provided in both sound and visuals, can enhance performance in terms of reaction time and accuracy if the sounds share features with and/or are congruent with the visual information (Blain et al., 2022). For example, we can provide task-relevant information through sound effects to help someone learn and navigate a new piece of software—this is referred to as the *multimedia effect* (Moreno and Mayer, 2000). For example, we might use error sounds to indicate to a user that they made a mistake, rather than use a visual pop-up, to reduce visual distraction and increase the speed with which they might learn. By offloading the warning onto sound, they can still be taking in a busy interface without added visual distractions.

3.5 Multimodality and Intersensory Integration

Multimodal interaction or integration refers to the fact that we don't experience our senses in isolation. We have many different *modalities* (or channels) of communication (figure 3.3), including:

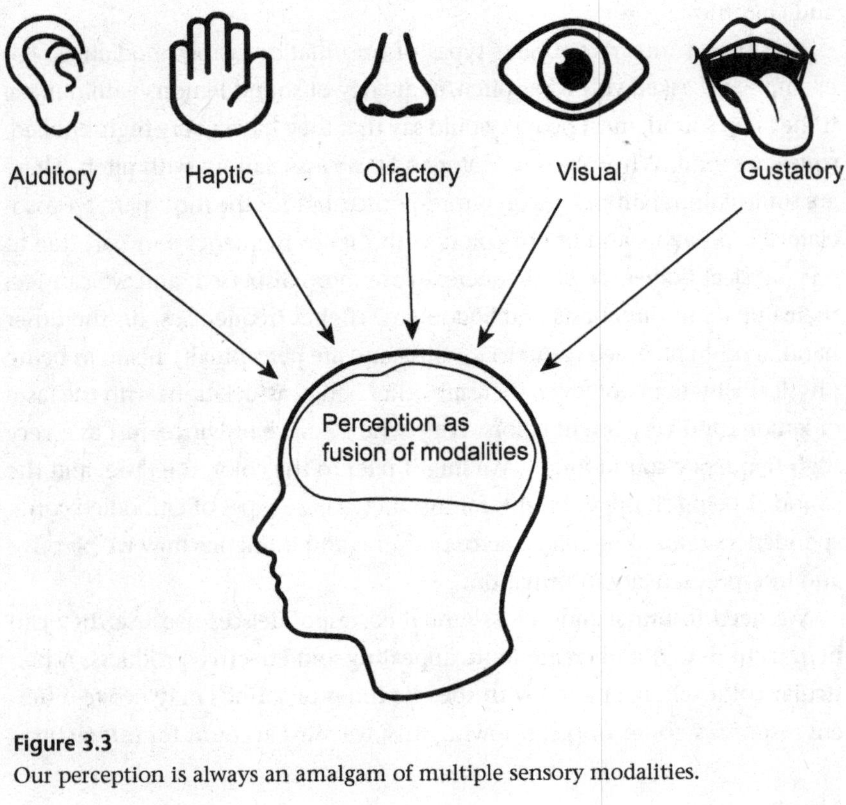

Figure 3.3
Our perception is always an amalgam of multiple sensory modalities.

- Visual: Information from the eyes, such as colors, shapes, text, and movement
- Auditory: Sounds and auditory cues, including speech and environmental noises
- Somatosensory/haptic: Touch, pressure, temperature, and pain sensations on the skin and within the body
- Kinetic: Physical movements, gestures, or actions
- Gustatory: Taste sensations derived from the tongue and mouth
- Olfactory: Smell sensations related to various odors in the environment.

Our brain combines information from our sensory modalities to create a unified and coherent perceptual experience in what is called *intersensory integration, cross-modal*, or *multimodal interactions*. In simple terms, our brain merges input from different senses, like sight, hearing, touch, taste, and smell, to form a more complete and accurate understanding of the world around us. This integration is a fundamental aspect of human perception and cognition.

Our brains integrate similar types of information across modalities. For example, if I asked you what pitch/frequency of sound lemons would make if they had sound, most people would say that they have a very high-pitched, squeaky sound. Why? We have color and taste associations with pitch. There are some cultural influences on our responses, but for the most part, we associate sharp flavors and bright colors with higher frequencies in part due to our physical bodies. Lower frequencies are more dispersed, and we can feel their impact in our chests and abdomens. Higher frequencies, on the other hand, tend to be much easier to localize, and are perceptually heard as being physically higher. Moreover, there are other bodily associations with the taste of lemons and very bright colors: They tend to make us wince, just as a very high frequency sound might. We might refer to the color, the taste, and the sound as being sharp, or bright, for instance. These types of embodied correspondences tend to be shared across cultures and influence how we perceive and interpret sensory information.

We need to understand cross-modal correspondences because they can help us to design and create more appealing and effective products: A particular color scheme paired with specific tones or sounds may evoke different responses, for example. Knowing this, we must account for interactions when we test our products: Any other sensory information can modulate our affective response to sound. This means if we're testing sound designs, we should test them in context of our product for many types of questions. However, we can also leverage this knowledge of intersensory integration to our advantage. When there is a deficit in one modality, another may be substituted: People with hearing impairments sometimes perceive sound with touch, for example (see below). For a thorough overview of sensory substitution, see Visell (2009). What follows is a summary of some research into sound's interaction with other modalities.

3.5.1 Sound and Taste

We often use metaphors to describe sounds in terms of taste (e.g., a sour note, a sweet melody) (Guetta and Loui, 2017). Some researchers and composers have attempted to create music specifically to leverage these correspondences (see Mesz et al., 2012). Studies have shown that taste words have musical correspondences, for example: Bitter is associated with lower pitches and continuous, legato music. Salty is characterized by a staccato rhythm, and sour is high-pitched, fast, and dissonant. Sweet is heard as consonant, slow, and soft.

Sound can influence our sense of taste (Velasco et al., 2023). Crispy foods when eaten tend to produce more high frequencies above 5 kHz, whereas crunchy foods produce more sounds of lower frequencies (1–2 kHz). Manipulating these frequencies can influence our perception of the crispness or crunchiness of foods we are eating (Spence, 2015). Packaging influences perceptions as well: The sound that the cork makes when removed from a wine bottle influenced ratings of wine (Spence and Wang, 2017), and the sound of a beer can or bottle opening likewise influenced perceptions of the taste of beer (Almiron et al., 2021).

The emotions expressed in music can also be transferred (in what is known as *sensation transference*) onto food being eaten. Music has also been shown to influence the perception of chocolate, with the chocolate being perceived to be softer when listening to positive-sounding music, and more intense with softer music (Reinoso-Carvalho et al., 2017). Music that is liked leads to perceptually sweeter tasting food, and the flavor of beer was rated differently depending on whether the listener liked or disliked the music being played (Reinoso-Carvalho et al., 2016).

3.5.2 Sound and Touch

One of the closest integrations of sound with another sensory input is with touch. Our auditory and tactile perceptions are very closely linked (Yau et al., 2009). Auditory cues frequently occur when we touch or interact with objects, and these sounds often convey potentially useful information regarding the nature of the objects with which we are interacting (Gaver, 1993). Touch has long been used as a means to decode auditory signals by the deaf and hard of hearing community. Tactile input also activates auditory areas of the brain: Even in adults with early-onset deafness, the auditory cortex is stimulated by vibrations similar to how they it would be with auditory feedback (Caetano, 2007). We also use touch-based terms to describe sound and music, such as soft rock or smooth jazz (Imschloss and Keuhnl, 2019).

It has been suggested that since we naturally evaluate the texture of many surfaces by stroking them with our fingertips, it follows that touch-related sounds will have a large influence on our perception of tactile texture (Guest et al., 2002). Research indicates that people can discriminate between different objects and/or surfaces on the basis of the sounds they make when touching those objects (Lederman, 1979).

A well-known study demonstrating the interaction between sound and touch is the parchment skin illusion. In this study, participants rubbed their hands together while listening to the friction sound over headphones. By manipulating aspects of the sound (high frequency content, or amplitude), it altered the perception of their hands as being more or less dry (Jousmäki and Hari, 1998). Guest et al. (2002) showed in a similar experiment that perception of the abrasiveness of paper could also be varied by altering frequency components (high frequencies resulted in the perception of greater roughness). Further experiments have shown that it is not only frequency components, but also other auditory manipulations that can influence the perceived texture (e.g., Spence and Zampini, 2006).

When it comes to things that we touch with our hands in particular, sound and touch are tightly intertwined. Ballas (2007) drew on studies of haptic-audio asynchrony to demonstrate that synchronous sound improves performance in a variety of tasks. Moreover, where sound is made asynchronous, even to a fraction of a second, there is a marked decrease in performance, which will influence perception of virtual environments if the timing is off even slightly (Adelstein et al., 2003). Studies have also shown that the addition of sound to a haptic interface can improve performance (Díaz et al., 2006).

When we are used to hearing actions or touch in association with a sound, the absence of that sound can significantly impair that experience. Sound can offer significant feedback and even successfully substitute for haptic feedback. Fernström, Bannon, and Brazil (2004) explored the impact of friction-like sound on texture perception, and found that users could detect the pseudo-haptic buttons almost as well as the haptic (real) ones.

An emerging field of *pseudo-haptics* has evolved, which leverages haptic illusions created by multimodal interactions, so that we can mimic the feel of real-life objects, particularly in the virtual space, using visuals and sound. Research has shown that when these auditory cues are presented in isolation, they can provide sufficient information for people to assess the size of objects and even what material they are made from (Freed, 1990). McGee et al. (2001) suggest that the use of sound can be employed to manipulate perception of texture in virtual devices, notably the perceived roughness, which has since been supported by studies into audio-haptic interactions using fMRI machines (Kassuba et al., 2013).

Other experiments have shown that sound can influence the stiffness perception of virtual objects in virtual environments (e.g., DiFranco et al., 1997). Cañadas-Quesada and Reyes-Lecuona (2006), for instance, explored sound's impact on the perceptual response to a haptic force-feedback device. Participants had to determine which of two virtual surfaces appeared to be stiffer. An auditory stimulus was introduced and shown to interfere with stiffness perception. Likewise, force can be simulated using sound (Watanabe and Yasumura, 2003). When combined with visuals, however, it was found that audio did not contribute to the perception of force itself, but instead was found to add to the perception of realness and presence (Biocca et al., 2002).

3.5.3 Sound and Action

There are some recent fascinating findings when it comes to the interaction of sound and our actions. Kinetic-based interactions with sounds, and the use of gestures as a metaphor for pitch changes in particular, have been researched (Lemaitre et al., 2017). In studies, participants have been asked to physically trace a sound in the air, performing free movements that mimic the sounds they hear, in what has been referred to as motormimetic sketching (Godøy et al., 2006): Commonalities are found between the respondents when it comes to gestures and sounds (Nymoen et al., 2013). In other words, we have an internalized bodily mimicry of sonic gestures that can be externalized through movement, but which normally remains entirely in our mind.

3.5.4 Sound and Image

The interactions between sound and image have been well studied from the perspective of psychology and neuroscience, as well as in areas of music and film/games studies, and to a lesser extent in product sound.

The idea of cross-modal correspondences between color and sound goes back to at least Pythagoras (Spence and Di Stefano, 2022). High frequency sounds will guide visual attention toward light-colored objects in a visual image, and the opposite is also true (Wang et al., 2020). In a retail study, for instance, high frequency music led customers to make more purchases from a light-colored shelf (Hagtvedt and Brasel, 2016). A series of studies have explored the influence of color on perceived loudness perception of various objects, including cars, trains, and coffee makers, with mixed results. One study found that women responded differently than men when it came to the influence of color, with the cross-modal effect more prominent in women (Papadakis et al., 2022).

Perhaps one of the most well-known of the cross-modal interaction effects is the *Bouba-Kiki effect*, along with its related studies (Köhler, 1947) (figure 3.4). This series of studies explored the idea that certain shapes of phonemes are associated with certain cross-modal properties: Bouba is heard as round and soft, while Kiki has sharp edges and straight lines. These associations have been demonstrated across cultures. An interesting recent study has demonstrated that the Bouba-Kiki effect also occurs when the visual domain is occluded (participants are blindfolded), and the participant is given shapes to touch, rather than see. In other words, sound-shape associations hold for the haptic-auditory associations as well as visual (Fryer et al., 2014).

Sound also has an influence on visual motion perception, including the perceived direction or trajectory of an object. The *ventriloquist effect* refers to the idea that we localize sound with an image, particularly if that image is moving, as when we see a ventriloquist's dummy *speak* even if the voice is coming from elsewhere (Choe et al., 1975). The movement of the dummy's

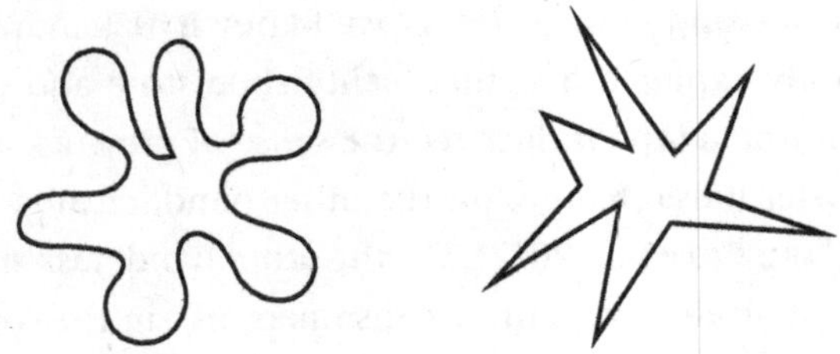

Figure 3.4
Bouba and Kiki: Shapes have correspondences to sounds, which holds for touch as well as visuals.

mouth allows for a suspension of disbelief as it draws our attention to the dummy, rather than its handler. Sound also influences perceived distance, with lower frequencies increasing the perceived distance (Sunaga, 2018).

While visuals most often influence the perception of audio, the inverse is also true. For example, beeping sounds can create the illusion of visual flashes (Shams et al., 2000). In another test, known as the motion-bounce illusion, two identical visual targets will be perceived as crossing through each other in the absence of sound cues, but when a brief sound is added at the moment that the targets interact, a bias perception toward the targets bouncing off each other occurs (Sekuler et al., 1997).

My own research with colleagues previously explored the impact of background sound on various aspects of visual fidelity (realism) perception in virtual environments. We found that background sound can impact our perception of high-fidelity imagery, with contextual sounds (those that belonged to the scene) leading to an increase in visual fidelity perception (Rojas et al., 2013). Hulusić et al. (2011) showed that sound effects allowed slow animations to be perceived as smoother than fast animations and that the addition of footstep sound effects to visual-based walking animations increased the animation smoothness perception in a virtual environment. Bonneel et al. (2010) observed that the visual level of detail is perceived to be higher as the auditory level of detail is increased. Such use of cross-modal illusions is useful where there is a need to drop frame rates or reduce visual fidelity. This in turn may allow us to increase the fidelity of the audio channel if the enhancement of visuals is economically or technically limited.

There has been a growing body of work into exploring cross-modal correspondences in virtual retail stores. For instance, the use of sound to enhance presence and immersion is well-known (Chandrasekera et al. 2015). One study explored a virtual reality shoe store and demonstrated that a slower music tempo influenced the sense of presence and intention to revisit. A faster music tempo, on the other hand, created more arousal and pleasure (Loureiro et al., 2021). On the other hand, task-irrelevant cues can have a negative effect on online consumers, as can the consumer's reasons for being there: Those who are just browsing, rather than looking for a specific item, will be more positive toward music. Online stores appear to show many of the same effects of music seen in in-person retail (Doucé et al., 2022).

3.6 Embodied Cognition

Embodied cognition is an influential theory that challenges traditional views of cognition by emphasizing the interconnectedness between the mind, body, and environment. As shown above, our perception of reality is significantly constructed through our sensory inputs. Embodied cognition theory suggests that our cognitive processes are not solely confined to the brain but deeply intertwined with our sensory interactions with the world around us. In essence, it suggests that our ability to think, reason, and perceive is shaped by the ways in which we engage with our surroundings through our physical senses. This means that our bodily sensations, gestures, and even the way we physically interact with objects can influence our cognitive processes, memory, and decision-making. According to embodied cognition theory, our cognitive processes use reactivations of sensorimotor states from our past experiences to understand the world. In other words, our knowledge is tied to the original neural state that occurred in our brain when we initially acquired the information or were exposed to that stimulus. Our cognition is therefore *embodied* because it is inextricably tied to our sensorimotor experience. In other words, our perception is always coupled with a mental reenactment of our physical, embodied experience.

Scientific evidence supporting the embodied cognition approach can be found in research into the mirror neuron system in the brain, through which a three-way mapping between sound, image, and action occurs. Research has shown that the same group of neurons will fire when either performing or observing an action, and so these are referred to as *mirror neurons*, although the extent to which they function in this manner is under debate. Embodied cognition theory can also help to explain multimodal interactions, since it shows that the way in which we interpret sounds is not solely an auditory process, but is inextricably linked to our physical state and rooted in our physical experience.

The concept that we mentally mimic the causality behind sounds we hear is not new to those who study music. Research has shown that we can recognize and feel the emotion conveyed by a performer when we listen to music, and this also results in physiological changes (blood pressure, heart rate, galvanic skin response) (Bresin and Friberg, 2001). An embodied cognition approach as to why this occurs suggests that we understand human-made sounds (including those of playing a musical instrument) in terms of our own experience of making similar sounds and movements. Our mental re-creation of the music causes a neuronal and motor-sensory response in ourselves that mimics the performer, and thus we are able to interpret the emotional inflections through our re-creation of the action. We therefore give meaning to sound in terms of mentally emulated actions (Leman, 2008). Put differently, we mentally and sometimes physically imitate the expressiveness of the action behind the sound, based on our prior experiences of creating sound (Cox, 2001).

3.7 Body Schema and Self-Produced Sound

Closely related to embodied cognition theory is the sense that we understand our own body in our environment, known as our *body schema*. Very young babies understand sounds they produce themselves versus sounds that are produced by other actions in their environment (Rochat, 1998). In other words, we have a physical experience of sounds that we produce ourselves that is different from other sounds in our environment. Our connection to sound is arguably much stronger when it comes to self-produced sounds (rather than externally produced sounds), because of the embodied connection between self and the sound.

Our sense of self is in part dictated by the sounds our body makes in interacting with the environment. People speak of being aware of their hearts beating in their chests at times of fear, or of breathing heavily when exercising. We have a continuous soundtrack of these sounds in our lives, referred to as our *sonic self-avatar* (Tajadura-Jiménez et al., 2017). When these sounds are removed, or we eliminate all external feedback, as in an anechoic chamber (a room designed to eliminate sound reflections), our sense of our body becomes altered (Noel et al., 2018). Moreover, modifying the action sounds we hear ourselves make has been shown to influence people's emotional response and bodily sensations (Tonetto et al., 2014). One study, for instance, found that changing auditory feedback when tapping a table could give the impression that a participant's arm is much longer than it actually was (Tajadura-Jiménez et al., 2012). Manipulating footstep sounds has recently been used to alter body illusions in people with eating disorders (Tajadura-Jiménez et al., 2022).

Our body schema can also extend beyond our physical body to incorporate non-corporeal objects. Recent applications of the rubber hand illusion have illustrated that sound works the same way to alter our sense of body schema (Radziun and Ehrsson, 2018). In this series of studies, a rubber hand is stroked in front of a participant, while simultaneously their own hand is blocked from view, but is also stroked at the same time. The participant's brain will take ownership of the rubber hand, such that any subsequent damage, such as pricking it with a needle, are felt by the participant. Research by Stanton and Spence (2020) has highlighted the role of auditory cues in influencing bodily and movement perception, underscoring the auditory component in the intersensory integration processes that contribute to body schema.

3.8 Anthropomorphism and Social Surrogacy

Anthropomorphism is the tendency to attribute human-like qualities, characteristics, motivations, emotions, and intentions to nonhuman entities, such as animals, objects, or even natural phenomena. Research has shown that the brains of those interacting with a nonhuman anthropomorphized object trigger the same mental processes as those involved in perceiving humans (Wan and Chen, 2021). We have a predisposition to anthropomorphize objects, particularly those objects with which we interact regularly, that have human-like physical characteristics, or that behave in "an apparently complex or intentional manner" (Heider and Simmel 1944).

Interactivity in particular contributes to anthropomorphism, notably through receiving feedback on our own actions, as in a movement or sound in response to something we do. Morse (1998) explains, "[A computer's] very capacity to give feedback and the immediacy of its response lends the quality of 'person' to what is a computational tool. This responsiveness allows it (and the virtual entities it displays) to pose or function as subjects—however partial, quasi, imaginary, and virtual—who are involved in the interactive exchange."

Product developers, who understand that anthropomorphism can benefit their designs, can give the product visual human characteristics (like cars having a face; Kühn et al., 2014), can give the product a name (such as the Henry series of vacuums), or can personify the brand itself (e.g., Mr. Clean; Wan and Aggarwal, 2015). Roughly one in three products sold to adults, and five in six sold to children, use some form of anthropomorphism (Triantos et al., 2016). Humanizing a product encourages us to treat a product as social, credible, and intelligent (Velasco et al., 2021). We trust anthropomorphized products more, and develop more attachment to those products (Velasco et al., 2021). Moreover, the emotion expressed by an anthropomorphized object can also influence our own emotions, just as in the social contagion theory of human interaction, where someone else smiling can induce a smiling response in ourselves (Yuan and Dennis, 2019).

When consumers lacked a sense of connectedness (e.g., were lonely), anthropomorphism helps make up these deficiencies (Chen et al., 2018). We have an increased tendency to anthropomorphize products when we feel lonely; seen for instance in the way Tom Hanks's character developed a strong attachment to a volleyball he named Wilson in the movie *Castaway*. The more socially isolated we feel, the more likely we are to seek connections with inanimate objects (Epley et al., 2007). Anthropomorphizing can lead to decreased uncertainty around a product, since there is a way of relating to the product, and it perhaps implies a sense of predictability or decision-making behind seemingly random results. For instance, gamblers who anthropomorphize slot machines will bid higher and have a reduced sense of risk (Yuan and Dennis, 2017).

While anthropomorphism is often associated with visual cues such as facial expressions, body language, movement, and gestures, sound also likely plays a role in this phenomenon. By employing sounds that mimic human-like vocalizations, laughter, or expressions of pain, creators can shape our emotional responses to nonhuman characters, making them more relatable. Even the mere presence of sound in animate objects such as robots or interfaces likely increases the tendency to anthropomorphize.

One study into listening to cute sounds and viewing cute images found that sound increased prosocial responses toward objects (Shin and Mattila, 2021). Another small study explored how participants assigned or perceived the attitude of a computer based on its beeps, finding that beeps of increasing intonation were perceived as disagreement, decreasing intonation was associated with agreement, and flat sounds with longer durations were interpreted as hesitation, showing that the beeps could be associated with assigning an attitude—and thereby anthropomorphizing—a simple computer (Komatsu, 2005).

Millicent Cooley (1998) demonstrated with a series of experiments of animated shapes that sound can especially help us to anthropomorphize an inanimate object and give it a sense of personality. Furthermore, the use of sound in media and technology plays a pivotal role in shaping our perception of anthropomorphism. In animation and film, even without using a human language, characters like *Wall-E,* or Toothless from *How to Train Your Dragon* are often given voices that mimic human speech patterns and emotions, making them relatable and emotionally engaging.

My own research looked at the role of adding sound to a variety of animations or objects and used several measures to test for anthropomorphism. We found that the two types of sound used (both cute vocal-like robot interface sounds and straight beep interface sounds) increased the tendency to anthropomorphize these objects over the same objects without sound (Collins et al., 2025). We tried to pick a boring, non-anthropomorphic object (a parking meter) and added sound to it. Although people tended to dislike our attempts to add cute sounds, they nevertheless had the intended effect! We also replicated an old animation of simple shapes by Heider and Simmel (1944): the original animation tested anthropomorphism of the shapes moving around the space, but had no sound. We found that women, and those from Africa, had stronger anthropomorphic tendencies than men or those from Western or Asian countries.

In addition to anthropomorphizing products, sound and music can also play a role in *social surrogacy*. This concept refers to how individuals use objects or media as a substitute for social interaction, companionship, or emotional connection when direct human interaction is limited or unavailable (Groarke et al., 2022). Studies have hypothesized that music can play this role, particularly when it comes to evoking nostalgia and reducing a sense of loneliness (Schäfer and Eerola, 2020). Listening to a favorite song or genre can evoke emotions and memories that simulate the presence of another person, creating a sense of connection even when physically alone (Schäfer and Eerola, 2020). Music can provide a stand-in for another person by expressing a mood or emotion and establishing a sense of connection to another (Davidson and Keene, 2019). The presence, or implied presence, of other people alters our experiences of a situation, changing our emotional reactions during the event, along with the subsequent memory (Monfardini et al., 2016).

Researchers have also found that sound effects can play this same role as music in terms of social surrogacy. While users can turn off sound in many mobile apps, the use of sound leads to increased social surrogacy, as well as higher app adoption intentions, higher willingness to pay, and increased likelihood to recommend the app (Poupis et al., 2021). The researchers suggested that since we know people are around us even outside our visual field by the sounds they are making, sound effects may imply the social presence of others and decrease loneliness.

3.9 Sound and Emotion

When it comes to sound, emotion is at the heart of our experience. We listen to music, for instance, to induce emotions, change emotions, release emotions, relieve stress, enjoy or comfort ourselves, or for other reasons (Juslin and Västfjäll, 2008). But what roles can sound and music play in emotion when it comes to products?

There is considerable debate in the research as to definitions and categories of emotion, affect, and mood. *Affect* is often used as a blanket term for all of these labels. I use the definitions provided by Västfjäll (2003), where emotion and affect refer to individual and immediate reactions to sound, and *mood* is, instead, a background state. For instance, a participant may come to an experiment with a positive, happy mood, but a single stimulus may evoke feelings of sadness, without impacting their overall mood. I can recognize a sound is sad, or even feel a moment of sadness myself, without it changing my underlying happiness that day, in other words. Studies have shown the important role of the underlying mood of the listener on perception of sound (Asutay et al., 2012).

There are many competing theories and models of emotion, and the subject is too complex to do justice to here (see, e.g., Barrett et al., 2007).

Contemporary theories of emotion often incorporate both cognitive and somatic (physical, embodied) elements, recognizing that emotions are complex phenomena influenced by both bodily responses and cognitive appraisals of situations. There are also many models that have been proposed to explain or categorize emotions, and these models have influenced both the scientific study of emotion as well as the methods used to study and test emotion.

The models of emotion that have been developed in the scientific community vary in terms of how emotions are categorized or described, and these influence the tests that have been developed to explore emotional response (see chapters 6 and 7). There is general agreement that we have a series of basic emotions and also more complex emotions, even though which emotions belong to which category are debated. Primary, or basic, emotions are relatively short and intense reactions to an event (real or imagined), encompassing a specific experience, cognitive and physiological change, and motivation to undertake an action important in terms of survival in physical or social environments (Janowski and Chełkowska-Zacharewicz, 2019). These appear very early in life, are perceived equally across cultures, and are associated with distinct patterns of physiological activation. These typically also involve some physically characteristic forms of expression—pinched eyebrows for anger, for instance (Cespedes-Guevara and Eerola, 2018). There is some debate about what these basic emotions are, or if they should be separated from other types of emotions, however. Moreover, there is disagreement as to whether some of them should even be considered emotions at all (such as surprise: see Susanto et al., 2020). Secondary emotions are more complex, and are evoked by memories and conscious thought.

The foremost competing models of understanding emotion are discrete models and dimensional models. *Discrete or categorical* models of emotion suggest that there are a distinct and finite number of basic emotions. This model, made famous by Paul Ekman, suggest that all secondary emotions are essentially derivatives or subsets of a few basic emotions, and that each has its own distinct biological and physiological markers that we can measure. Variations on Ekman's model include Plutchik's wheel of emotions (figure 3.5). The validity of discrete emotion models has been questioned in recent decades, and has been supplanted in the scientific literature by dimensional models (Barrett et al. 2007).

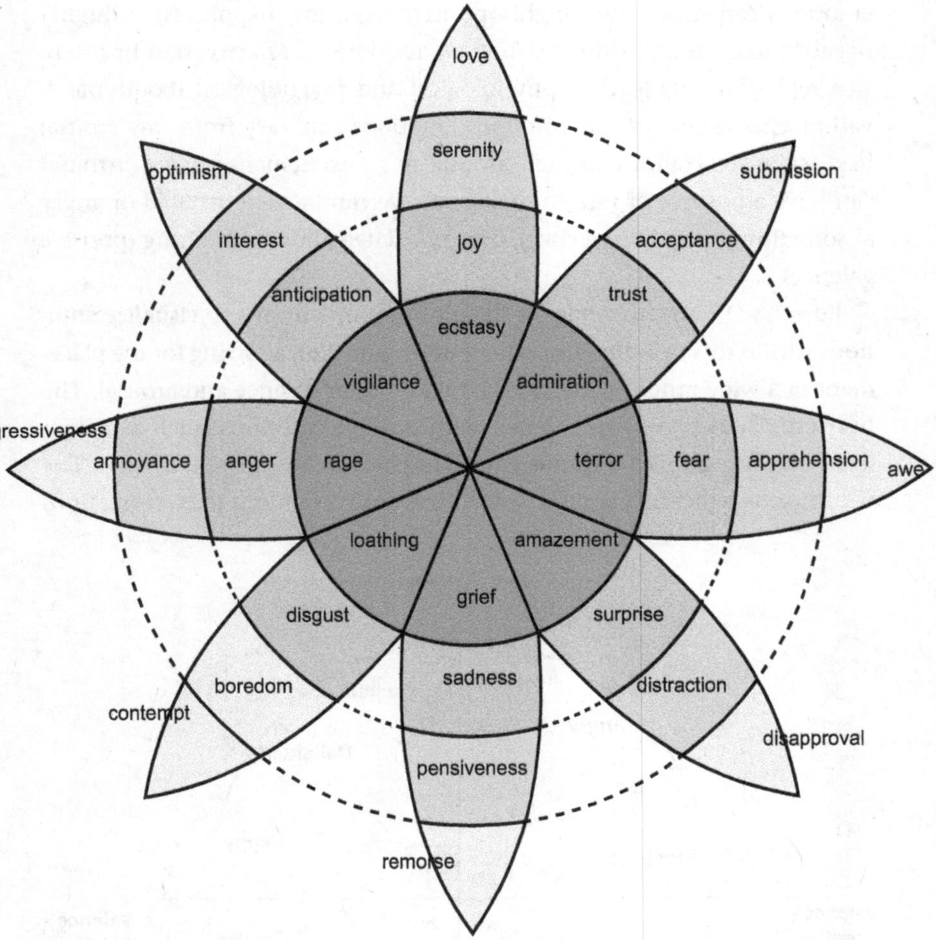

Figure 3.5
Plutchik's wheel of emotions (adapted from Wikipedia-Machine Elf 1735, https://en.wikipedia.org/wiki/Robert_Plutchik#/media/File:Plutchik-wheel.svg).

Dimensional models of emotion, on the other hand, suggest that emotions can be better understood by considering two or more continuous dimensions along which emotions vary. The most widely recognized and studied dimensional model of emotion is the circumplex model, which proposes that emotions can be plotted on two primary dimensions: valence and arousal.

Valence refers to the pleasantness or unpleasantness of an emotion, reflecting happiness and satisfaction. In simple terms, the valence of emotions can range from highly positive (e.g., joy, happiness) to highly negative (e.g., anger, sadness). *Arousal,* also known as activation or intensity, represents the level of physiological and psychological (bodily) activation associated with an emotion. Emotions can vary from low arousal (e.g., calm, relaxation) to high arousal (e.g., excitement, anger). Arousal can have a positive or negative valence—we can be very irritated or angry at something (negative valence), or very excited about something (positive valence).

Russell's (1980) circumplex is the most popular means to visualize emotions within this two-dimensional model (figure 3.6), allowing for the placement of a wide range of emotions based on their valence and arousal. The horizontal axis represents valence, with positive emotions (such as happiness) on the right and negative emotions (such as sadness) on the left. The vertical axis represents arousal, with low arousal emotions (e.g., relaxation)

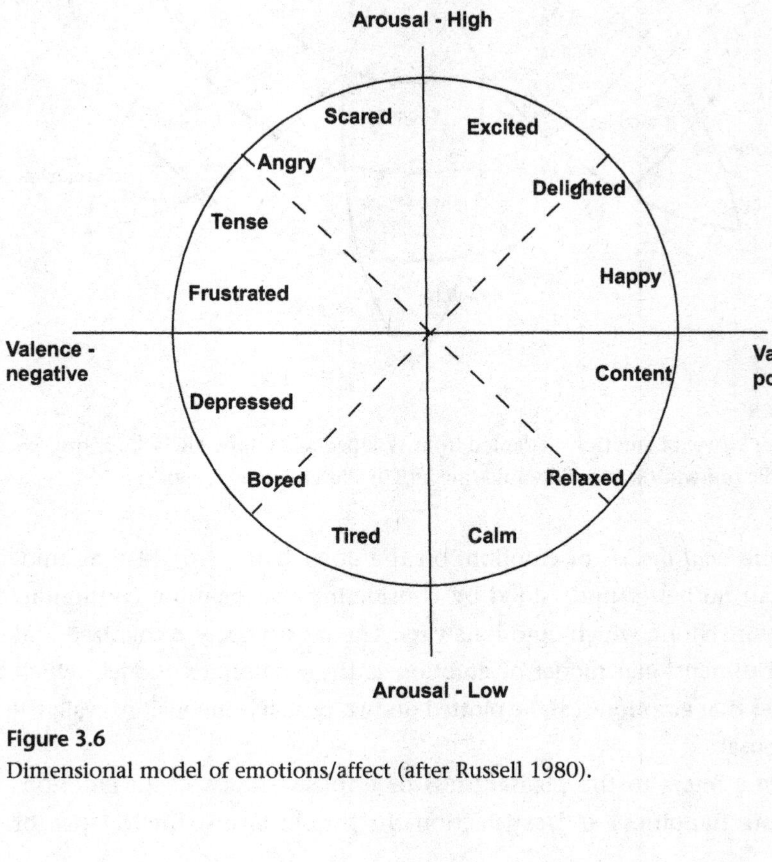

Figure 3.6
Dimensional model of emotions/affect (after Russell 1980).

at the bottom and high arousal emotions at the top. Different emotions are positioned within this circular space based on their specific combinations of valence and arousal. For example, joy might be located in the upper-right quadrant, indicating high positive valence and high arousal. Sadness might be located in the lower-left quadrant, indicating negative valence and low arousal.

Researchers have expanded upon this basic two-dimensional model by adding additional dimensions to capture other facets of emotions, such as pleasantness, tension, potency-control (the sense of control or power associated with an emotion), or unpredictability (Fontaine et al., 2007). Moreover, we could also add motor activation (approach/avoid/neither) as an element when it comes to basic emotions (Heilman, 1997). Tension in particular is of interest because music is often a process of tension and release, and resolution of tension through an expectancy effect is part of the pleasure of music (van der Zwaag et al., 2011). Expectancy violations increase arousal and induce surprise, and our expectation of what we are about to hear plays a role in our emotional response to sound, even triggering different areas of our brains.

A meta-analysis of sixty-one studies determined that *musical* emotions are not the same as our everyday emotions in terms of their intensity or physiological changes (Cespedes-Guevara and Eerola, 2018). Musical emotions do not prepare us to take any action, and are reactive, rather than proactive. Researchers describe, instead, a set of music-specific emotions, which overlap with what some have described as *aesthetic* emotions (Scherer et al., 2013), including feelings such as solemnity, fascination, admiration, wonder, and being moved. Moreover, some emotions are stronger or more profound when elicited by music, while others are very difficult to evoke with music (e.g., jealousy; see van der Zwaag et al., 2011). However, another analysis suggests that there is plenty of evidence, including self-reported feelings, physiological responses, and activation of brain areas associated with musical emotions, that some researchers feel that musical emotions should be classified the same as our other emotions (Juslin et al., 2008). Zentner et al. (2008) introduced a widely embraced classification of musical emotional responses in formulating the Geneva Emotional Music Scale, consolidating emotions into nine categories: wonder, transcendence, tenderness, nostalgia, peacefulness, power, joyful activation, tension, and sadness.

There is considerable debate in the literature between those researchers who believe music evokes a genuine emotional response (the emotivist position) and those who believe we merely perceive emotions expressed by a piece of music (the cognitivist position) (Lundqvist et al., 2008). Studies have shown there is in fact a difference between recognized and induced emotion: We may recognize a singer is angry, without feeling anger ourselves, for instance. Nevertheless, many people report a change in emotion invoked by music, and the ways we use music to pump us up at the gym or calm us down at the end of a day are clear ways that show music helps to alter the ways in which we feel. Whether these effects carry over to just sound effects or not is unknown, however.

An alternative approach to thinking about sound and music's influence on emotion is the multifactorial process model, which suggests that the observation or imagination of expressions of the musicians elicits muscular and neural mimicry, which then causes an emotional contagion, aligning with mirror neuron theory (Cespedes-Guevara and Dibben, 2022). For instance, we can often feel the music being expressed by a performer when listening to a piece of music, and the theory suggests we mentally imitate that feeling behind the performance. A similar effect is seen with sound effects, where we mentally reenact the action behind the causality of the sound.

Recent research suggests that there are other contributing factors to our emotional response to music, including listening with others, which increases emotional response (Juslin et al., 2008). Moreover, our baseline levels of empathy and even our genetics can influence our responses, or at least the degree of responses felt (Sittler et al., 2019). Studies have also shown that men and women experience music differently in terms of their perception of emotional expression, accuracy in decoding the emotion intended, facial responses to facial expressions and to auditory stimuli, and finger temperature response to music stimuli, with women having more pronounced responses than men (Lundqvist et al., 2008).

3.10 Conclusions and Take-Aways

In this chapter I've presented a lot of research that stems from areas of psychology, neuroscience, and sound/music studies. As described, the work in this area is mostly very recent and ongoing, and there is much to learn about sound. The following is a summary of the key points:

- Sound can evoke emotions and influence user behavior. Designers can use this knowledge to create interfaces that elicit specific emotional responses, such as calmness or excitement, to enhance user experience.
- The impact of sound on consumer behavior is influenced by factors such as the complexity of the sound/music, personality traits, and the context in which the sound is presented.
- Sound can influence human behavior through classical conditioning, seen in Pavlov's dog experiment. Conditioning, as seen in games and gamified apps, reinforces behaviors through positive or negative sound feedback.
- Our past experiences and associations with sounds can shape our perception and interpretation of them, influencing our emotional responses. Familiarity with music and sounds can increase liking and engagement, but overexposure can lead to decreased liking.
- Understanding how sensory inputs shape cognition can help designers create more immersive and effective user interfaces and experiences. For example, incorporating sound feedback that mimics natural gestures can enhance the user's sense of presence and engagement in virtual environments.
- By incorporating sounds that mimic human vocalizations or emotional expressions, designers can make nonhuman interfaces more relatable and emotionally engaging. This can enhance user experience by fostering a sense of connection and empathy toward the product, making the product seem more credible, intelligent, and trustworthy. By associating specific sounds with product actions or responses, designers can influence users' perceptions and attachment to the product.
- Sound can serve as a social surrogate by evoking emotions and memories that simulate the presence of another person. Designers can use sound in interfaces to create a sense of companionship or emotional connection, particularly in situations where direct human interaction is limited.
- Anthropomorphizing products through sound branding, such as using specific sounds or music to represent a brand, can create a strong emotional connection with users and enhance brand loyalty. Designers

can use sound to create a more immersive and comforting user experience, particularly in applications where users may be seeking social connection.

- Consider how different sensory inputs can be combined to enhance user experience. By understanding how actions are coupled with sound, designers can create interfaces that facilitate learning and skill development. For example, using sound feedback to indicate correct or incorrect actions in a training application can help users learn new skills more effectively. Additional modalities can also lead to unexpected perceptual effects and illusions, and change the meaning of sound.
- We must be aware that sound influences how a user learns and retains information. Presenting irrelevant sound can lead to lower performance. On the other hand, adding task-relevant sound can increase learning speed and reaction times.
- Sound can evoke vivid mental imagery, which can closely emulate actual perception and influence emotion and behavior. Studies suggest that mental imagery is a key mechanism behind emotional responses to sound, and it can be multimodal, involving more than just visual imagery. This imagery can be conscious or unconscious, and it has been shown to impact perception, memory, and learning. Sound effects and music will increase mental imagery in most audiences, with more content resulting in greater imagery. Highly evocative imagery will likely increase recognition and memory for products.

It's important to remember that sound and music are not universal languages. As I described, in my own study we found differences between the responses of men and women, as well as Westerners (North America and Europe) when compared with Africans. There are many distinctions between cultures in how they hear sound—sometimes, positive and negative sounds can be completely reversed. A study of the Khowar and Kalash tribes of northwest Pakistan perceived positive and negative stimuli opposite to British listeners, for example (Lahdelma et al., 2021). Consonance and dissonance have also been shown to be culturally specific (Lahdema et al., 2022), and timbre preferences differ by culture (Kim et al., 2016). Preferences for type of interface sounds also vary by culture (Jeon et al., 2015). If your product is designed for a specific market, and you wish to invoke a specific effect, be sure to take into account demographic differences and test on wide groups of participants where feasible.

The research into the psychology of sound is much newer than research into music. As such, while we can likely extrapolate some of the findings about music to sound effects as well, we simply don't have the research yet to back up some of the assertions and assumptions. Testing is, therefore, the only way to find out how your sound is going to work.

4 Designing Sound

This chapter will provide a brief overview of the process of designing a sound: not the nitty-gritty of how sound designers select, record, create, or manipulate sounds, but the process of deciding what sounds are needed, and the exploration and prototyping of the sounds as part of a design team, to share with stakeholders. In this way, the rest of the team may understand what is involved in designing and discussing sound, in order to be able to contribute and have fruitful conversations with the sound designers.

It is very common to meet executives at companies who do not understand the time that it takes to make effective sound design. Today, companies accept that their logo or their website can take months to design and is better left to professionals, but they still treat sound as if it's something that can be knocked out by anyone with a piece of audio software or a keyboard (or, perhaps worse, access to a text-to-sound AI generator). And due in part to a lack of understanding of what sound can bring to a product, the executives or even the design team will often accept whatever they are given as long as it seems "good enough" on first listen. Many companies fail to properly resource their sound teams (or have a sound team in the first place). At the other extreme, top companies recognize the value of sound: It's been reported that BMW has over fifty sound designers on their team, all working to perfect the sound of their cars.[1] They even hired Hollywood composer Hans Zimmer to compose their electric engine sounds. It takes expertise, expense, and time to do sound right.

Budgeting for sound design can be difficult, depending on the needs of the product and the complexities involved. Speaking very generally, the sound budget should at least match that of the product's visual design. Recording and producing high quality sound requires equipment, and mixing and mastering can cost thousands of dollars per track of music, if custom music is involved. Creating an audio brand such as a sonic logo can take many months and often requires extensive trial and error. Don't leave sound to the end when it comes to your budget, as it should be a carefully considered part of your overall design and marketing budgets.

4.1 The Sound Team: Whom Do I Need to Hire for This?

Having an in-house sound team is not feasible for most smaller companies, so often the work is contracted out. In addition to a composer and/or sound designer, a consultant or employee (such as a creative director) who can liaise between the visual design team, marketing team, and the sound contractors is also useful in ensuring that the sound you get meets your goals as quickly and as efficiently as possible.

A *sound designer* creates and designs sounds. They may have training in music composition but are primarily experts in sound design. On the other hand, while composers *can* produce sound effects, they might lack experience in sound design. The two arts are related, but different, and sound design is not usually taught as part of music programs in schools. Depending on your needs and budget, you might seek someone who can handle both sound design and composition, or hire separate specialists.

Sound designers may be responsible, for example, for interface sound effects in an app or for the seat belt warning signal in a car. A composer, on the other hand, may be responsible for a series of songs for on-hold music, advertising music, or in-store music. When it comes to branding, these roles can be somewhat blurred in the creation of a sonic logo. In an effort to create a distinct logo, the role of the sounds themselves becomes especially important (their timbre, particularly), and thus the creator of a sonic logo is often tasked with both crafting the sounds and constructing them into a pattern or motif that makes sense as a longer composition.

There are other roles involved in music creation in particular that are of relevance, as they impact budgets. Sometimes composers or sound designers may take on these roles, but there are also technical roles that require specialized training, experience, and intuition. *Producer* roles can vary by project, but often producers will help composers to make decisions about how something should be recorded, what effects to use, what instruments to use, whether to work with a mixer (see below), how to decide what mix to use, or other decisions. There may also be *orchestrators* and *arrangers*, who decide on the instrumentation and write instrument parts, although this is usually for longer-form orchestral music.

Depending on your goals, such as if you plan on recording any live instruments or orchestral music, there will also be performers and recording engineers for those sessions, in addition to the costs of renting the recording space. *Recording engineers* are a specialized role, and are involved in the setup of microphones in recording live instruments as well as potentially mixing. There are also typically *sound mixers or engineers*, who decide on the overall mix of the music—what effects to use to bring out the clarity of the instrumentation. *Mastering engineers* make final adjustments to any recorded music and prepare it for release. This may be handled by a mixer, but often is another level of the mixing process.

Creative directors and/or consultants may oversee the entire process, contract out the team, sit in on meetings, and help to direct the team, ensuring a unified brand. Since sound work is not usually done in-house, and many companies have little knowledge of where to start or how to describe what they are looking for, a consultant or director can step in and provide the guidance necessary to the sound team, and liaise with the corporate side. Much of my own consulting work has been in this role, liaising between the sound team and the design and marketing teams and facilitating the communication necessary to discuss the sound needs and decide what user experience testing should be done.

The cheapest budgets may get away with one person with a good music and sound design background and some high-end software packages. On the other end of the spectrum, it's possible to go all out and end up with massive teams of people required to coordinate. Recording a full orchestra, for instance, can easily involve over a hundred people.

4.2 Why Can't I Just Use AI?

If you're suddenly panicking at the thought of an expensive sound task, such as crafting a high-end sonic logo for your brand, or needing an orchestra for your advertisement, you may be tempted to just go to a sound library or use artificial intelligence. After all, with just a few key words in the prompts, an AI can spit out a sound effect or an entire orchestral piece. And sometimes, it may even be possible to use AI to effectively solve some sound problems. However, AI-generated content often lacks the depth of human emotion and nuance that a good sound designer or composer can bring. An effective user experience is about creating an emotional connection with your users, evoking specific feelings and memories through sound. Humans understand the cultural contexts, the emotional subtleties—and your brand—much better than an AI can. Moreover, generative AI operates on patterns and data from existing works, which can lead to dull, lifeless output that lacks the originality that you need to stand out in a crowded market. Humans can think outside the box, experiment with new sounds, and push the boundaries of what is possible, ensuring that you get something unique. Your product or brand's sonic identity should reflect your unique values, culture, and personality, not sound like most other examples out there.

If you're not convinced, on your next sound project hire a human and run user tests alongside an AI-composed example to compare. My own work has shown that the majority of users can nearly instantly hear the difference between AI and human-composed music (Collins and Manji, 2024). Another caveat is that the legal quagmire of that generative AI and the training models they require may mean that you end up having to redo your sound in the future if you've used AI, in order to satisfy some new copyright or legal requirement. Having to start from scratch can result in even more expense.

4.3 Talking to Stakeholders About Sound

Perhaps the most important need for many companies is simply selling executives on the concept that it is worthwhile investing in sound. Not only do UX researchers still face resistance to UX research at many companies, but sound is doubly difficult for many companies to justify spending money and time on, in part because they don't understand it, and partly due to old company cultures that are slow to adapt to the changing needs of sensory experience design. Opposition to change in the organizational culture and a lack of understanding UX and sound can be the most difficult objections to overcome, even more than budgets!

Sound is typically under-resourced compared with other areas of design and marketing. While companies will think nothing of investing $100,000 on a high-quality camera to shoot their videos for their social media, they can't seem to understand the value of spending $10,000 on a good microphone kit. It's rare that all stakeholders have the vocabulary or knowledge to talk about sound. It's also common that they may not immediately see the value of sound and what it can bring to the company or product. You may be on a sound team and need a higher budget, or on a design team that needs to convince executives to double the design budget to include sound. You will undoubtedly find yourself at some point having to justify the expense for sound.

In *Articulating Design Decisions*, Tom Greever claims there are three things every design needs to be successful (Greever, 2010, p. 10):

- It solves a problem.
- It's easy for users.
- It's supported by everyone.

To communicate with stakeholders and ensure that it is *supported by everyone*, then, we have to be able to answer three broad questions:

- What problem does it solve (in other words, why do we need sound?)
- How does or will the sound impact the user?
- Why is it better than the alternative(s) (in other words, how will sound improve the product in ways nothing else will)?

Answering these questions of course requires some useful user data. If you have the data, present the results of user experience testing validating the choices along with your resource request. Without data, it's useful to bring in some of the arguments about how sound can work on a subconscious level to influence user emotion and behavior.

When it comes to presenting your research and your request to stakeholders, there are several questions you should be asking in advance (Brown et al., 2016): Who needs to be included in the presentation? Does the entire sound team need to show up to be able to answer questions or is everyone on the same page and can they all be represented by one or two people? Which of the stakeholders should be present? Who makes the final decisions? Who is going to fund or support the project? Whom do you need to get on your side and what are their backgrounds, goals, and obstacles? You need a different presentation for the engineers who need to know what kind of sound-making device needs to be included than you would for marketing executives who need to know how to sell your ideas to get the budgets. What does the audience know about sound? What potential objections might stakeholders have? Play devil's advocate and anticipate any objections that decision-makers may have about the sound.

When discussing sound with stakeholders who lack a background in sound or music, it's important to communicate effectively, using clear and relatable language. The following are some strategies to help you convey your ideas about sound to non-experts.

Start with the basics: Begin by explaining fundamental concepts in simple terms. Clearly outline the goals of the sound design. Are you aiming for brand consistency, emotional engagement, usability enhancement, behavior modification, or something else? Sell the benefits of sound. Starting small—suggesting a smaller, low-stakes project with perhaps just a few sounds—can also be a way to begin a discussion about sound at larger companies.

Provide visual aids, analogies and anecdotes: Visual aids, such as diagrams, charts, or infographics, can help illustrate your points and make the information more understandable and comfortable or familiar to those who may be talking about sound for the first time. Relate sound concepts to experiences stakeholders might be familiar with. For instance, you can discuss how sound in movies creates tension similar to suspenseful moments in a book. Use anecdotes, industry benchmarks, or other real-world examples to illustrate how sound can shape perceptions and influence reactions. If a picture is worth a thousand words, some sound examples are worth a thousand pictures (Schifferstein and Desmet, 2008).

Focus on impact: Instead of delving into technical details, emphasize the impact of sound. Explain how sound can evoke emotions, set a mood, or enhance experiences. Play sounds and have them talk about how it makes them feel. What data do you have to support your work? Executives love data, and the reporting of solid data (see chapter 9) is a great way to get them on board. Highlight the immediate benefits of having sound, and the negative impacts of not implementing sound. Likewise, seek out research to back up statements. Are there research papers that will give you the data you need to show what you want to show if you don't already have some data gathered? For instance, if you are arguing that you will sell more products if you include sound, can you back that up with some research and/or data that stakeholders can grasp?

Use everyday language and familiarity: Avoid jargon and technical terms as much as possible. Use simple language that anyone can understand.

Consider how best to emotionally connect: Do you know their favorite band, or a movie they love that you can reference the soundtrack for? Studies have shown that metaphors are useful and commonplace to describe sound, particularly when it comes to a sound's timbre, so use terms like *warmth* and *roundness*, and *bright* and *sharp* (Rosi et al., 2023). Use multimodal metaphors to help them feel comfortable discussing sound.

Connect to goals: Frame sound in terms of its relevance to the stakeholders' goals. Explain how sound can contribute to the project's success, whether it's a marketing campaign, a product launch, or user experience of a product.

Provide scenarios and context: Explain the different scenarios and contexts in which the sounds could be experienced. For example, are there specific environments where the sounds should be more prominent or subtle?

Demonstrate, don't lecture (show, don't tell): If possible, provide sound demonstrations to let stakeholders hear the differences you're talking about. Early prototypes can be used, but be cautious—just as some people can't see the good bones in a house that has been poorly staged for sale, some people can't imagine a higher fidelity prototype and why you would need to invest more money when you just played a *perfectly good* sound!

Be patient: Remember that sound might be unfamiliar territory for them, so patience is key. Be willing and prepared to repeat or rephrase your explanations as needed.

Make a cost–benefit analysis: How might sound factor into a purchasing decision for your user? What is the anticipated business impact? What are your competitor's solutions and how will you improve what you're doing to pitch the competitive advantages?

Use A/B comparisons: A favorite old trick is to play a scene from a movie where the soundtrack has been changed. I like to use *Jaws*, which has such an iconic motif, and then swap out the original music for an Astor Piazzolla tango. The change in perception of the scene is undeniable, as it changes from tension and horror to a comical spoof. It's a very quick way to share the idea that so much of the emotion of what we experience is driven by sound.

Some cases that can be greatly aided by a successful listening demo are:

- Showcasing a new sound feature or sound quality improvement: Let's say users of your software were getting annoyed that they were missing important messages from the team. Show how adding sound will draw their attention to the missing information, and provide data illustrating how the addition of sound improved the user's experience.

- Showing progress: As discussed above, great sound takes time, and you will likely have to present progress reports. Be sure to include some prototypes if not finalized sounds. Listening demonstrations are the best way to show a difference: between you and your competitors, between you and a previous iteration of your own product, between the sound versus no sound, and other comparisons. Make sure that the difference is strong enough to support your argument by running at least a pilot study or informal test on different people in advance of the presentation.
- Demonstrating a proof of concept or a mock-up: Showing your product with and without sound is an obvious way to help to illustrate the value of sound and what it can bring. Be sure to be ready to counter the annoyance-over-time rebuttals with the science that shows the value of including sound (Brown et al., 2016).

4.4 Building a Brief for the Sound Team

The sound team will be grateful to receive guidance on what a client is searching for in terms of the sound. A clear brief is no substitute for a director who understands the brand and products and can communicate this to the team, however. An outside consultant can also perform this liaison work between sound designers and the rest of the company, and help to formulate the brief for the sound team, if there is nobody well-versed in sound design in the creative team.

Brand attributes: A brief should include detailed information about the brand overall, including visual brand (color palettes, fonts, and visual references of products that are under the scope of the brief). Keywords, especially related to the emotional attributes of a brand, are very helpful: For instance, if your brand is serious versus playful, luxurious and opulent versus simple and functional, or your company is trying to express concepts like freedom or beauty. The more information you have about the character of the company, the brand attributes, and the products involved, the better. Sound is so important to the subconscious feel of a product. Be sure to understand what emotions or mental imagery you want to evoke, and the emotional context of your product's purchase and use.

Use-cases: Sound can be a detriment to a product if not used appropriately. Begin by mapping out use-cases of the product and where sound might help the user experience. If formative user research has been undertaken, any relevant findings should be included in the brief—for instance, if users have difficulty understanding that they've taken an action, a sound could be used to draw their attention, so information of that nature is useful to the designer. What problems are you solving with sound? Where is sound expected to be used? (This may change, depending on the product and the sound designer's creativity and any user research undertaken.) Describe the typical use of the product, the situations and environments where it might be used. Provide any competitor analysis or user experience research that has been undertaken. See chapter 5 for further guidance here.

Detail the use scenarios of your product for the sound designer, so they might determine what roles sound might play. A function analysis will indicate not just the obvious functions of a product (i.e., the actions and tasks that a product does), but also the less obvious functions. The insights that can be gathered in a function analysis can also help the sound designer to prioritize sounds. Tools such as function trees or function maps can help to illustrate the details of processes and common features (Michalakoudis et al., 2017).

Understand technical limitations and/or known constraints: If you are working with a physical object, understand the technical limitations of sound playback on the device. Small speakers work better with higher frequency sounds, but these can sound tinny and cheap. Be sure to understand your known constraints—are you creating sound for a mobile phone application, or are you expecting the sound to be played through high-end desktop speakers?

Prototype: Depending on your expertise, you may wish to prototype a palette of sounds that you feel reflect the brand and/or product (see section 4.5, "Prototyping," below).

Existing guidelines: If the company already has a sonic signature that defines it, for instance, a sonic logo, then that information should be included in the brief with sound files and details of the sound:

- What notes are used (melody)
- What instruments/synthesizer patches (VSTs, plugins, etc.) are used and what settings they use
- What processing/effects are used and what settings are used in those effects

- How instruments are being played (if relevant; for instance, fingerstyle guitar is different from picked guitar, and even within fingerstyle, there are different ways of strumming, such as with fingerpicks or with bare fingers; in addition, the strings on the guitar may make a difference, as does the style of guitar). Also include any specific instrumental articulations such as particular emphases or note accents, playing legato versus staccato, and so on.
- Recording guidelines (including miking distance and position, or other factors)
- Production guidelines (mixing styles, production notes).

A brief may also include a competitor's sound analysis, or just references to the competitor's sounds and what your team likes or dislikes about the competitor's sounds.

File naming conventions and formats: Naming conventions are useful to include, particularly for larger companies that may be creating a sonic library for various uses. As well, include requirements as far as audio file formats and deliverables (e.g., deliver as *stems*, or as a final mixed product? Deliver as WAV or compress as MP3? Master to what dB or loudness units (LUFS)?)

Follow-up work: When you have sounds from the sound designer, engage in user testing and feedback sessions, and iterate on the sound design as necessary. Outline any plans for testing the sounds, and plans for iterations. Review and update the sounds regularly to ensure that they stay aligned with the brand or product changes. See section 4.7.3, "Future Proofing Your Sound Design," below.

Use of sound in the product: When the sound is implemented in your product, develop guidelines for your sonic brand (see below), creating a charter about where and how and what sounds should be used, so anyone building anything related to the product will have a guide as to what are acceptable and preferred uses of the sound.

4.5 Prototyping

Sound prototyping involves creating and testing sonic elements (sounds and music cues) in the early stages of product development, ensuring that the sonic feedback provided to users aligns with the intended user experience goals. Prototyping provides a platform for stakeholders to provide feedback and suggestions, which inform design refinements and help identify potential issues before a product is fully developed. Nevertheless, as has already been discussed, prototyping sound has some complications, in that many people can't hear a difference between a low fidelity prototype and a final version, and may become attached to earlier versions. Note also that testing different types of prototypes can yield different results. Moreover, most sound designers don't really prototype sounds, and aren't familiar with the concept. They usually just go on intuition and deliver high quality sounds from the start, so you may want to consider prototypes as just early versions of sounds that haven't yet been tested or tweaked.

Most people are familiar with visual sketches and mock-ups, but how do you *sketch* a sound? A graphic artist can quickly map out a rough sketch of a logo or product's appearance, but when it comes to sound, our sketching tools are a lot less obvious. Sketching is a necessary part of the creative process, and exploring ideas through sketching is invaluable. However, communicating these ideas is much more difficult when it comes to sound. We can certainly leverage cross-modal correspondences and draw out what we are trying to talk about, or try to capture what we mean with a gesture (Lemaitre et al., 2017), but there are also sonic tools we can use. There have been various academic attempts at sound sketching toolkits (e.g., Delle Monache and Rocchesso, 2019), and the use of tools will depend on the designers involved and the type of sound being created, but we can summarize them roughly as reference material, vocal sketching, Foley, and synthesizer tools to create sound for video, virtual, or physical prototypes.

4.5.1 Reference Material

It's often useful to find reference materials to use as examples of the kinds of sound you are looking for. In film, these are called *temp tracks* or reference tracks, and are temporary music compositions from other films that are used to convey a mood or feel that the director is going for. For example, it's probable that "The Imperial March" in *Star Wars* (1977) was based on Holst's "Mars: Bringer of War" from *The Planets*. There are also many video collections online that show film music that was clearly inspired by other film music. When it comes to product sound design, it's also possible to use reference material if you have a particular sound in mind. For instance, a mock-up can be made using an online stock sound library such as Freesound, Epidemic Sound, or Splice, or even new AI tools that allow you to prompt a generator with some keywords. These sound effects may or may not be usable in an actual project, depending on the licensing involved, so any information about where you obtained them along with licensing information is always important to track. However, sometimes over time we grow accustomed to temp files and can become attached to them, such that when the new sounds are created, the new sounds feel out of place. It is important, then, to recognize that temp files should only be used for a short period of time and as a way to communicate ideas quickly. Competitor references can also be collected for this purpose: Collections of sounds that competitors use for similar products can begin as a jumping-off point for discussions of what you like and don't like about an existing sound design in a similar product.

4.5.2 Vocal Sketching

Vocal mimicry is an important part of human development. Indeed, our very language is littered with vocal sketches, in the form of onomatopoeia. We are quite adept at attempting to imitate or create sounds with our mouths, known as vocal sketching (Monache and Rocchesso, 2019). The voice is our quickest, cheapest means to communicate sound ideas, although vocal sketching has its limitations. For instance, the voice is *monophonic*—that is, we can usually make only one sound at a time. When it comes to timbre, we are also somewhat limited in terms of the types of sounds we can create, and complex sounds and percussive sounds can be quite difficult. Duration is limited to the breath, although this typically isn't a detriment for most product sounds. Nevertheless, vocal sketching is useful for communicating ideas like melodic contour or a short theme or individual sound effect, and is very diverse, with some practice. Video games like *Hidden Folks* (sound design by Martin Kvale) have even been created where all of the sounds for the game are created through the voice.

Vocal sketching can be as simple as humming a melody or creating vocal percussion to establish a rhythmic foundation. Vocal sketching is not about achieving perfect pitch or professional singing; instead, it focuses on capturing the essence of the musical idea and exploring its potential. Vocal sketching offers several benefits in the music creation process:

- Rapid idea generation: The voice is always readily available, and vocal sketching allows us to quickly generate a multitude of musical ideas without the need for instruments or software.

- Freedom and expressiveness: Singing or vocalizing ideas can be an expressive and uninhibited way to communicate emotions and artistic intentions. Vocally mimicking the sound, moreover, can help us to feel in our own bodies how a sound might communicate emotion (see section 3.6, "Embodied Cognition," above, for the importance of our body in interpreting sound).
- Easy collaboration: Vocal sketches can be easily shared and understood by others, facilitating collaboration and feedback during the early stages of a project.
- Melodic exploration: Vocal sketching allows us to experiment with various melodies, harmonies, and rhythm combinations to find the most compelling musical phrases.
- Refinement and iteration: By recording vocal sketches, we can listen to, analyze, and refine ideas, leading to the development of more sophisticated compositions.
- Inspiration: Vocal sketching can be an excellent tool for overcoming creative blocks and sparking new ideas, as it encourages a more intuitive and spontaneous approach to music creation.

For example, Hug and Pfaff (2019) describe the sketching process for the contour of a robot voice to express emotion and draw out the contour of the sound. By drawing the tonality, or melodic contour, we can direct sound designers to adapt their sounds; see figure 4.1.

Researchers have found that vocal imitations were as effective as verbalizations in communicating identifiable sounds, but that when sounds were abstract or not clearly identifiable, the use of vocal sketches really excelled over verbally attempting to describe sound. They describe that retaining

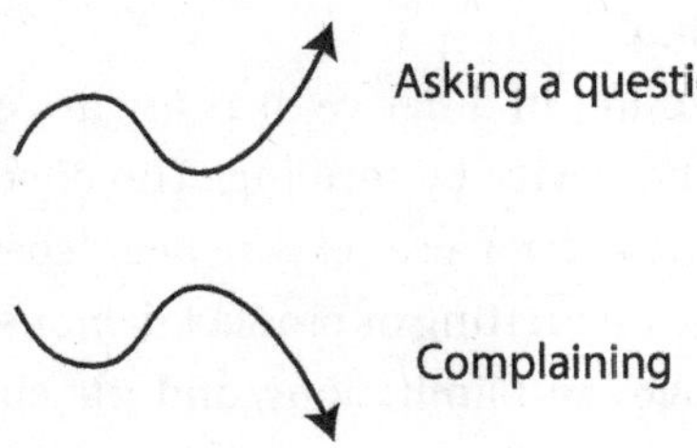

Figure 4.1
Drawing the gesture of a sound, adapted from Hug and Pfaff (2019).

spectral and temporal information (e.g., timing, contour) is crucial when imitating sound (Lemaitre et al., 2016).

Vocal sketching is just one type of quick sketching of sound, although it's perhaps the fastest since we all generally have the instrument required at our disposal. Other authors have proposed, for instance, rapid sketching with *low-fi* sound: unpolished sound effects that capture an essence (Marentakis, 2023), such as a melodic contour or gesture.

4.5.3 Foley Sound Prototyping

Foley is a sound design technique used in filmmaking and other media productions to create and enhance sound effects that complement the visuals on screen. Named after its pioneer, Jack Foley, a sound editor in the early twentieth century, Foley involves the recording and reproduction of everyday sounds, actions, and movements to match the actions of characters and objects depicted in a film or video. Using various props and materials, Foley artists perform and record footsteps, cloth movements, object interactions, and other sound effects in sync with the visuals. For example, a Foley artist might recreate the sound of footsteps on different surfaces by walking on various materials like gravel, sand, or wood. The goal is to produce natural and realistic sounds that match the visuals and immerse the audience in the film's world.

We can use a Foley pit to similarly prototype some types of sounds, particularly when it comes to percussive sounds, impact sounds, or physical sounds of falling, crushing, flapping, breaking, rubbing, and other sounds. We can sync them to videos of the product for rapid prototyping. However, Foley is limited without further manipulation in software (Hug and Kemper, 2014).

4.5.4 Video Prototyping

Before implementing sound in a device, it is usually quicker to overlay a sound on a video of the device performing. The effectiveness of a video compared with a real device for user experience depends on the specific context and goals of the user testing or product demonstration. Both methods have their advantages and limitations, and the choice between them should be based on the particular use case. My research into the effectiveness of video versus live object demos for testing sound's meanings found some differences in interpretation of sounds between the two presentation modes, with sounds presented in person being rated more positively (Collins et al., 2024). However, for the most part, the results were similar enough to suit the purpose of prototype testing.

Creating a video demonstration of a product can be more cost-effective and accessible than producing real devices for each user. With a video, you can ensure that all users experience the same demonstration, reducing variations in testing conditions. Videos can be quickly edited and refined to showcase the product's best features and eliminate any irrelevant content, and video demonstrations can be easily shared and distributed to a broader audience than a hardware prototype, allowing for remote testing and reaching more potential users.

However, users cannot directly interact with the product in a video, which may limit their ability to fully explore its functionalities. Actual devices allow users to interact with the product in a real-world context, exploring all features and functionalities, providing more accurate and authentic feedback. A video cannot replicate the tactile or sensory experience of using a real device, which may affect the user's perception and feedback. Users can physically touch, feel, and handle a physical prototype, enhancing their understanding and perception of the product. On the other hand, depending on the product, users' perceptions and feedback may also be influenced by the novelty or excitement of interacting with a physical device, leading to biased responses. Moreover, providing real devices for user testing can be expensive and logistically challenging, especially in remote or large-scale testing scenarios.

4.5.5 Sonic Moodboards

Moodboards are commonly used in a variety of design practices. Most of us are familiar with the visual moodboards used by designers and clients to share ideas and discuss textures, colors, and imagery. I have been using moodboards as a sound design teaching tool and find that they are a very useful tool to explore emotion and mood as related to sound. There are several approaches that we explore in my workshops: Users may create a collage of sounds related to express a particular mood or idea, essentially, individual, clickable sounds on a PowerPoint slide or web page. A second approach is to create a short (twenty- to thirty-second) sonic collage that changes over time, incorporating and layering several elements. This latter approach is more effective in trying to capture and express a particular mood sonically.

Moodboards can be useful not only to talk to stakeholders about the types of sounds or general mood being aimed for, but also to explore how the interaction of sounds may influence the overall mood, for testing emotional/mood communication abilities of the sound, and for other purposes. For instance, in my class we test a student's abilities to create a mood by having other students try to guess what was being expressed through the choice of sounds. Moodboards can stick to plain, unaltered sounds or can use various effects and processing to further push the sounds into a particular mood. For instance, a little distortion on a sound can make a regular sound appear more aggressive or angry, whereas reverb can add warmth and depth.

4.5.6 Physical Prototyping

When it comes to interactions, it may be important to test how well a sound responds to a user's action; in these instances, physical prototyping may be necessary, and for some products, it's just necessary to interact with the physical device to get the full effect. Physical prototyping with sound design is typically the last stage in product sound design prototyping, since as described above prototypes can be more expensive to test. It's possible to create Wizard of Oz style prototype sketches sometimes (where aspects of the prototype are "faked" for testing). On one project testing robot sounds, for instance, we built a little 3D-printed backpack for the robot to carry a small Bluetooth speaker, rather than build out the audio capabilities of the robot itself. Using a physical prototype more accurately represents the real final product, which is to say, there is a higher degree of fidelity in this type of prototype. But temporary sensors or interactions can also be built quickly into physical objects using an Arduino microcontroller board or a Raspberry Pi–type computer and software.

4.5.7 Cooperative Design Practice

A recent trend in user experience has been to use co-design or participatory design workshops for designers and their users, emphasizing the active involvement and collaboration of stakeholders throughout the design process, with the aim to create products that better serve the needs of their users. Co-design might be undertaken in the early stages of design, for instance, in asking users about their desired auditory experience, in exploring the types of sounds they find pleasant or irritating, and in other ways. Sonic moodboards can be created, and various types of prototyping can take place with users providing input on their preferences (see, e.g., Monache and Rocchesso, 2016).

4.6 Style Guidelines for Sound Design

Just as a graphic charter or style guide details the parameters of the brand's visual image (e.g., logo, fonts, color guides), a sonic style guide should also be created. Depending on the size of the company, this may extend to a sonic library that can be used in a variety of different ways, or may be just a simple sonic logo with guidelines as to where, how, and when to use it. The guidelines will help to maintain consistency. This ensures that all brand-related materials, from advertisements to social media posts, have a cohesive and recognizable sound.

A guideline should also outline how the brand's sound should be used, specifying where it should be used, what variations are acceptable in which situations, and what not to do with the sound to maintain its integrity. You may need to provide different versions of the sonic logo for various contexts, lengths, or emotions. Identify the various touchpoints where the sonic elements will be used—advertisements, on-hold music, animations, or other areas. Provide guidelines on appropriate volume levels for different scenarios to ensure a comfortable and pleasant listening experience. If applicable, the guidelines might provide instructions on how the brand's sonic identity should be applied to physical products, packaging, retail outlets, hold music, and other applications.

Guidelines should include legal information, such as trademark usage and copyright notices, to ensure the brand's intellectual property rights are protected. Outline the legal aspects of using sound elements and music, ensuring compliance with copyright laws. You may have licensed a popular song for certain uses, for instance, but not for others, so be mindful of where, when, and how licensed material can be used.

Brand guidelines often include templates and examples of various implementations to serve as practical references for designers and content creators. Define the preferred musical instruments, genres, or musical elements that resonate with the brand. Specify the overall musical styles that align with the brand's identity. Describe the emotions or moods that the brand's sonic elements should evoke—for example, energetic, sophisticated, friendly, or innovative. Explain how these emotional attributes might change based on context, such as a customer support hold line versus a product launch video.

Provide details on the sounds themselves, such as which sounds, instruments, notes, registers, tempos, effects, mixing protocols, and file types and naming conventions are used. Be sure to have very specific details on the effects and the settings used within the effects, so if your sound designer changes, the new one can recreate the original if necessary. Provide guidance on adapting the sonic elements for different cultures and regions, if applicable.

4.7 Challenges and Considerations in Implementing Sound

Incorporating sound into a product, and designing those sounds, is not without challenges and additional considerations to be thought through early in the design process. Done well, sound should add to, not subtract from, a product's experience. As such, it's necessary first and foremost not to annoy anyone using the product, or those around them. I suspect we've all been annoyed by product sounds at some point—a child's toy, someone's mobile phone, the microwave. I once worked in an open office for a tech company where someone had their email configured to have Homer Simpson yell every time a new email came in: funny for the first ten minutes or so, but after several months? Not so funny. Conor O'Sullivan, Google's lead sound designer, states that the biggest challenge for a sound designer is to "play sound for people that don't necessarily always want to hear sound or don't want at least to notice these sounds all the time."[2] Sound is not always needed and can distract users. Beyond annoyance, there may be considerations about regulatory requirements (e.g., noise emission standards, consumer health and safety regulations about volume, electromagnetic and radio frequency interference), accessibility requirements, industry standards (such as for alert levels), cultural and social considerations, privacy laws, technical restrictions, and more: There is a lot to consider when designing sound for a product!

4.7.1 Annoying Audio: Where and When to Use and Not to Use Sound

Once we learn about sound and what it can bring to a product, it can be tempting to put sound everywhere. But knowing when and where to use—or not use—sound can be critical to the product's success. Asking questions about which areas of your product need sound is probably the first consideration you should have about the product's sound design. Understanding the behavior of users and how often they might hear that sound is also critical. I want my washing machine to play a sound to tell me it has finished its wash, but I don't want a notification that every cycle in that wash has been completed. I want my car to tell me if I forgot to buckle my seat belt, but I don't want it to shout at me that the backpack I put on the passenger seat needs a seat belt (because of an overly sensitive sensor that the dealer assured me could not be rectified).

Consider the length of sound and its regularity: a more frequently heard sound should be shorter and less intense. Notifications that something has happened or is about to happen that the user needs to be aware of are also very useful. A simple, short notification is enough.

However, even these sounds can be felt to be intrusive. My friend, for instance, once posted on her Facebook page, "While nearly all my friends are creating sounds and music for TV and movie or game soundtracks, I have been *getting rid* of sound in my kitchen: My microwave *beep*. I have a newer microwave and the dang beep is cracking the crystal. So I took off the microwave's shell, found the offending electronic component, covered both sides of it with craft foam . . . and continued the hopeful attenuation with a part of a Magic Rub grey block eraser, taped in place with 2-sided carpet tape. . . . This appliance does not drive me nuts with noise now." Someone on Reddit, in an unrelated thread, agreed: "Microwaves are the rudest of all appliances! They beep every 30 seconds or so when they are done cooking. Come on, microwave! I didn't forget I was hungry, my hands are full, wait your turn!"[3]

Responses to the Facebook post are telling about the lack of usability:

- "Hmmm . . . is there not a setting to turn off the beep? Instructions may be buried deep in the owners manual." Response: "I looked! And I just wanted to turn it *down*, not *off*."
- "I think manufacturers think we put something in the microwave and immediately go shoot hoops in the driveway."

- "I have an air fryer which I swear has a beep that's 70dB. It's *insanely* loud for an appliance beep, and there's no volume adjustment, nor the ability to turn it off. Pretty sure the neighbors know whenever I use that device. Your solution here is giving me ideas. Edit: I informally measured it from about a foot away. Obnoxious" [picture of a readout of 73.1 decibels].
- "My teeny 700 watt microwave beeps sooo quietly it reminds me of a drug dealer going *psssst!* trying to get my attention at a police station. I have found day-old decaf tea I'd forgotten about because I didn't hear any beeping."
- "I have a modern toaster that makes a very unnecessary beep-beep-beep when the toast is done. ALL toasters for the past several decades have made a sound when finished . . . ka-chunk! No beep beep needed! Idiots."

There was a lot of engagement on the post, and many people had opinions on their appliance sounds! Some wanted theirs louder, others quieter. A simple user option to adjust settings to their circumstances and preferences would have solved these frustrations.

It's important to consider the context in which the sounds will be heard. A product that is used in public should make less sound than one in private, as anyone who has ever had to deal with someone else's keyboard clicks on a mobile phone in public can attest. On the other hand, sounds can be incredibly useful when visibility is reduced, or the eyes are on other tasks. Anyone who drives a car with a manual transmission probably shifts gears based on the engine sound without even thinking about it. In this case, constant sonic feedback about a changing state is important to the successful and safe running of the product.

Sounds can also become annoying over time, as my example about the Homer Simpson email notification can attest, so the repetition of sounds should be taken into consideration. Some sounds are great the first fifty or so times we hear them and then can become annoying (see the discussion of the Wundt curve), particularly tonal sounds. The most annoying sounds tend to be in the frequency range of 2,000 to 5,000 Hz, the range most likely to trigger emotional responses (Kumar et al., 2012). Repetitiveness can be mitigated by making sounds less likely to grab your attention, but this may also defeat the purpose of the sound.

Natural sounds, or sounds that fit with the sonic environment, are likely to be less annoying than unnatural sounds. A bird chirping outside my window could be ignored for days, for instance, but the regular repetition of a bird sound inside my office at one-minute intervals would quickly become annoying, first, because the regularity of it means my brain will perceive the pattern, but also because birds don't belong in my office! If there were a slight variation to the chirp—it varies in pitch a little each time, in length, or in the interval time—the annoyance would likely be less because it would sound more natural, even though it's out of context.

Until we were able to record sounds, no sounds were ever repeated exactly the same way twice, and our brains are very adept at noticing changes in sound. From an evolutionary perspective, being able to listen for changes in the sounds around us was very important for survival. Now, it is entirely possible to repeat a sound ad nauseum, and while we sometimes need to listen for survival (when crossing a road, for instance, and hearing the cross-walk beep signals), it's far less imperative. Slight variations can help to overcome the repetition, but nonetheless, if the sound becomes *too* variable it can change its meaning to us, and might lose its purpose. A careful balance must be struck. Anecdotally, user interface sounds do not need to be a direct one-to-one mapping, but can exist within a narrow range of a particular sonic framework. This technique is common to video game sounds today: In the past, the same two sounds would repeat for a character's footsteps: one, two, one, two. The repetition could be ignored by many players, but for others, the artificiality of the footstep sounds took them out of the game. Today, footstep sounds can be randomly played back from a bucket of options, and various effects and filters ensure they are played back differently each time (or, alternatively, they are procedurally generated using granular synthesis or artificial intelligence, so entirely new sounds are created each step!). As one sound designer explained to me, "I'd knock on a table at different places (edge, center, side, etc.) and say 'listen: each of those sounds are different and unique, but you still recognize them as *knuckles on tabletop*. As long as the sounds are of a type, I'd say *play from bucket* is just as effective as *play same sound each time*."[4]

Sometimes a two-stage signal is necessary, with an initial signal to let the listener know that the important notification is about to follow, such as an announcement system that provides a short beep first to signal to listen. As seen with the sonic logos discussed earlier, the listener often doesn't notice the signal sound, and only hears the logo notes. A signal to listen is sometimes important to start a significant notification.

It goes without saying, as well, that in most cases there should be the option to turn off sounds. Sounds should generally be matched to the relative urgency of the information they are indicating. If sounds are an important safety feature (like the seat belt warning sound), then the user perhaps shouldn't be able to control them (and regulations may dictate what is allowed); however, in most cases, being able to shut off, change volume, or customize sound should be a feature, particularly when it comes to products that are likely to be used in public, where sound-making could cause embarrassment or annoyance.

It is also important to coordinate with other modalities where possible, such as haptics or visuals on a phone interface, for times when the sound would need to be or is desired to be turned off, and for those with hearing loss. While it is tempting to eliminate such redundancies, pairing across modalities can help to reinforce conditioning effects, and also provide information and feedback when sound is not available. A backup modality is also useful for when users may simply miss the sound because the environment is too loud, because they are distracted, because the product is obscured sonically in some way (a phone is buried in a handbag, for instance), or when users are hard of hearing.

4.7.2 Sound Design and Accessibility

Accessible design means all users can enjoy your product. Why cut out some of your potential market unnecessarily? Consider users who might have hearing impairments, and be sure to provide alternatives such as visual cues or haptic feedback in addition to sound cues, or if you have a critical warning sound, make sure it's not a single frequency in case your users have lost that one frequency. Hearing loss in the speech range affects an estimated one in five adults, and one in three are affected by loss in high frequency ranges (Ramage-Morin et al., 2019). Most people with hearing loss (an estimated 77 percent) are not even aware that they have some hearing loss (reported in Ramage-Morin et al., 2019). Be sure to incorporate accessibility into your design, in terms of both the sounds selected and the ways that they are used and played back.

Incorporate accessibility into your user testing—be sure to test on older adults and people who wear hearing aids, for instance, particularly if your product is likely to be used by these groups. Allowing customization means users can select sounds that are appropriate to their needs.

When it comes to web-based products, which are also often applicable to other software, the W3C Web Accessibility Initiative offers guidelines for sounds. These include, for example, using pause/stop options for any automatically playing sound that is longer than three seconds, and ideally not automatically playing any sound, particularly on pages where there is text, since the sounds interfere with screen readers used by the visually impaired.

On the other hand, sound can also be used as an augmentation for the visually impaired. Consider designing sounds in a way that visually impaired people can quickly learn your sound design scheme, such as by using sonic hierarchies, to better use and understand your product.

Beyond physical disabilities, accessible design should also consider the impact of sound on those who are neurodivergent. Autistic individuals are more sensitive to sound, and can have an increased auditory perceptual capacity, but also tend to experience decreased tolerance to sound (known as hyperacusis, misophonia, and phonophobia—see Remington and Fairnie, 2017; Williams et al., 2021). On the other hand, other neurodivergent people can experience the opposite effect, and are hyposensitive, enjoying loud sounds. Working with diverse audiences in your user research can help to mitigate some of the issues you may encounter with product sound, and offering customization options is always a great way to ensure the product appeals to a diverse audience.

4.7.3 Future-Proofing Your Sound Design

Future-proofing your sound design involves creating a design that remains relevant and effective over time, even as technology, trends, and user expectations evolve. Sound preferences evolve along with visual preferences, and you don't want to end up with the lens flare equivalent of audio! The following are some strategies to help you future-proof your sound design.

Focus on timeless elements: Rather than chasing current trends, prioritize elements that have a timeless quality. Classic, simple, and universally appealing sounds are less likely to become outdated quickly.

Modularity and flexibility: Design your sound assets in a modular way, allowing you to easily update or replace individual elements as needed without overhauling the entire sound design. For example, think about keeping the melody or rhythm and changing the instrumentation and timbre.

High-quality original recordings: If possible, create original sound recordings that are of high quality. These recordings can be tailored to your brand's unique needs and won't be as subject to the trends of stock sound libraries, which can sound out of date more quickly.

Adaptability across platforms: Design sounds that work well across different platforms and devices. This can help ensure that your sounds translate effectively as technology evolves.

Consider minimalism: Minimalistic sound design tends to have a more timeless quality. It's less likely to clash with evolving visual design trends, and can be easier for users to process. An exception here is sonic logos, which can be more complex.

User-centered design: Keep the user at the center of your sound design decisions. Base your choices on user feedback and behavior to ensure the sounds remain relevant and effective for your target audience.

Future trends research: Stay informed about emerging technology trends, user behavior shifts, and design best-practices. Anticipating changes can help you adapt your sound design proactively. Sound effects don't follow trends as much as music or visuals do, but music especially is prone to feeling dated after a few years. An example here is the typing sounds on a mobile phone: As described earlier, young people view these as a signifier of being old.

Regular updates and iterations: While you aim for a future-proof design, understand that some updates will still be necessary. Plan for regular reviews and iterations of your sound design to keep it fresh and aligned with evolving needs.

Documentation and asset management: Maintain clear documentation of your sound design choices, including the rationale behind certain decisions. This documentation can help guide future updates and iterations.

Design for scalability: If your user interface is likely to expand with additional features or content, design your sound scheme to accommodate growth without causing disruptions.

Stay open to evolution: While aiming for future-proofing, also stay open to the possibility that some aspects might need to change significantly to adapt to unforeseen changes.

4.8 Conclusions and Take-Aways

This chapter introduced some important considerations before beginning your sound design process. We can summarize some key points in a series of dos and don'ts.

1. Don't let the sound annoy anyone using your product, or the people around them, unless it is a critical safety feature. Your users should in most cases have control over the sound—if not to change the sound, then at least to turn down the volume or shut it off. Sounds should enhance the user experience without being intrusive. They should complement the visual elements rather than overshadowing them. Provide users with control over sound settings, including volume levels and the option to mute sounds. Respect users' choices and avoid surprising them with loud or unexpected sounds.
2. Account for repetition. Sometimes, a sound is great once, maybe twice, but after the thirtieth time you want to pick up the product and throw it out the window. There are many ways sound designers can handle repetition, and this should be considered when planning your sounds. Sounds should not be used for actions that are highly repetitive. Natural sounds (i.e., sounds we would hear in the world), muted tones, and musical sounds are generally felt to be less intrusive than synthesized, noisy sounds. Familiar and predictable sounds, like the ticking of a clock, are felt to be less intrusive as well, since we learn to ignore them. You can introduce variation into your sound design by generating sounds in real time, and programming small variations if your technology affords this kind of variability. Variability will make the repetition less noticeable, since our brains are very adept at picking out patterns. Sounds should not repeat ad nauseum in most circumstances, but should allow for variation, or irregular repetition, where possible.
3. Users should have control over the volume of sound where possible. Customization should be considered. Often users may want to turn down (or turn off) sound, particularly in public. But sometimes they may also want to turn it up! My washing machine's sound is too quiet because the machine is in my basement, so if I put a load in, I have to leave my music off if I want to hear when the laundry is finished. If I had control over the volume, I could turn it up a bit and I would appreciate it much more.
4. Don't use sound that serves no purpose. Sound should serve a purpose in your product. It shouldn't be there just because you can put sound there. I remember the early days of the World Wide Web, when people were first able to embed music in their sites. It seemed, for a time, that

everyone felt their page should have a soundtrack and embedded MIDI files into their sites, resulting in users having to manually mute pages. Thankfully, that practice has disappeared. Understanding why you are using sound is central to using it!

5. Don't use sound that will not reproduce well on your device. Your sound should sound good, which is to say, it should reproduce well on your target device. If you have a great sound that sounds perfect in your studio and take it out and pipe it through a 25-cent speaker, does it still sound good? If you put a lot of effort into the bass frequencies but have a tiny speaker that can't reproduce sounds at those frequencies, what is the point of it? Make sure what you produce can be reproduced well. Likewise, if you spend a lot of time spatializing a nice stereo sound, delicately planning out panning effects, but you have only one speaker on your device, what was the point? If your user interface spans multiple platforms (e.g., web, mobile, desktop), ensure that the sound design remains consistent across these platforms to provide a unified experience.
6. Reflect the tone of your brand. You don't want heavy metal music if you're selling luxury rings, unless your target audience is the rock stars who perform that music. Likewise, don't use sound effects that don't reflect your brand. Use experts who know what sound is communicating, and run everything through user tests to confirm. Understand your target audience and their preferences. Different user groups might have varying reactions to certain types of sounds, so tailor your sound choices to match the user context. "If it's designed right," stated O'Sullivan, "the user won't really think too much about that, and it'll just become a part of the overall fabric and the overall experience of engaging with that product" (Google Design, 2018).
7. Employ clear, distinctive sounds with a consistency and coherence throughout the design. Sounds should be clear and easily distinguishable from each other. Avoid using similar sounds for different actions, as this can lead to confusion. Maintain consistency in sound design throughout the product. This helps users develop a mental association between specific sounds and actions, making the interface more intuitive. Make sure that the sounds used are relevant to the context of the action. For example, a button press should have a different sound than a notification alert.

8. Be mindful of the cognitive load of your user, and of the cognitive load that sound can add to the user's experience. Too many different sounds can be overwhelming and detract from the usability of the user interface. The presence of irrelevant or excessive auditory stimuli can increase cognitive load by demanding attention and mental resources, thereby diverting focus from the task at hand. On the other hand, carefully designed and relevant auditory cues can aid in information processing and task performance. For instance, in educational settings, background music that complements the learning material can enhance engagement and memory retention, while chaotic or distracting sounds can hinder comprehension. Effective management of sound's influence on cognitive load is essential in various contexts, such as user interface design, learning environments, and workplaces, to ensure optimal cognitive processing and a seamless user experience.
9. Understand the needs of software developers managing data and memory allocations. If your user interface is for a mobile app, consider the potential impact of sound on data usage. Avoid using high bitrate, lengthy audio files that might consume a lot of your user's data (especially if you have a travel app that the user may be downloading abroad). Another obvious example here is sounds in video games, where a developer is not culling (removing) sounds that have already played, and may lead to constant calls on the device's memory, creating a lag in the playback or other playback issues. If you have no idea what I'm talking about, just ask your developers to consider sound and performance issues.
10. Be culturally sensitive and be aware of diverse audiences. Be aware that certain sounds might carry cultural meanings or associations that could be misinterpreted or offensive in some contexts. Choose sounds that are neutral or universally understood. Avoid stereotyping by gender, race, sexuality, or other areas when you are creating your sounds. Be aware of accessibility needs relating to sound.
11. Own your own sounds. It should go without saying that sound falls under the same copyright laws as images, and you shouldn't use anyone's sounds that you haven't paid the rights for. Ensure that the sounds you use are either original or properly licensed to avoid legal issues related to copyright infringement. Believe it or not, professionals can pick out music parts, or sounds, stolen from elsewhere. That means the originator of the source of your theft is likely to come across it and recognize it.

5 Evaluating Product Sound: Asking the Right Questions

Whether you're developing a new design, implementing a rebrand, testing an innovation, or seeking to fine-tune an existing feature, a user study is a compass that can guide your decisions with real-world insights. At its core, a user study is a structured exploration that seeks to answer questions about your design and about your users' behaviors, thoughts, emotions, and expectations.

Well-executed user research must begin with formulating your research questions. What is the purpose of your user study? What are you seeking to learn? What data do you need? What don't you know that would be useful to know? What problems are you trying to solve? Starting with your broader purpose will help you to narrow the focus of the specific questions you seek to answer. Think through how you intend to use the results as well—the results needed will help you to prioritize your questions and decide what is manageable for a study or how many studies you may need to run.

Here, I am only presenting questions specific to sound. Product designers should also, of course, be conducting general user experience research beyond sound, but which may also include sound, for instance, relating to the pain points of your customers, their experience with existing products, experience with your brand, and other factors. There are many books on the fundamentals of user experience research that can be helpful if you are new to general user experience research (e.g., Pannafino and McNeil, 2017; Travis and Hodgson, 2019). As such, rather than repeat this existing work, I want to focus here on where sound intersects with the user experience.

This chapter will focus on helping you to think through your study's aims, and to formulate potential sound questions you might seek to answer with user testing. Specifically, the focus here is on deciding what you want to measure in your research. The following chapters will discuss the setup of experiments and recruiting and running sound tests, as well as more details about the methods involved. Many of the questions presented here relate to other questions, so there is some overlap between these categories and questions. Which questions you want to answer will be different at different stages of the design process, but it's always good to think ahead and discuss these with the whole design team right from the start.

Think through the stakeholders on your project, to determine who should be involved in formulating and defining your UX research questions. It's also helpful to gather any existing data that you have on your users (or intended users), along with data about your brand, and review these as a starting point. Research starts from the questions, not from the methods.

User tests should be considered when you are introducing a new product or new version of a product, when you want to know what features you should be prioritizing in your design, when you are entering a new market or new area as a company, before you launch a product, and after you launch a product (to identify problems and successes). Consider, first, what your overarching aims will be of this particular test: Is it to learn about your users, to learn about usability issues, to get data from users about preferences and needs, to support iterative design and get feedback on your progress, and/or to validate design decisions, improve user satisfaction, and confirm a product meets market needs or enhances your brand's perception?

5.1 Formative Design Questions

Formative design research questions are those questions that you may want to ask at the earliest stages of the design process, before you even have a prototype, or at an early prototyping stage. As discussed, sound is often left until the end of the design process, and doing that means we miss out on many opportunities to make sound an integral part of the overall design and the overall user experience.

5.1.1 Is the Inclusion of Sound Necessary? What Functions Should Sound Serve?

Any product should begin with asking, does it need sound? What problem is being solved with sound? Sound can be a feature, but it can also be distracting, annoying, or stressful. Why are we including sound, and is sound a necessary part of our product? In the early days of the web in the mid to late 1990s, many websites automatically played background music on their sites. This trend, fortunately, faded out for the most part, since most users found sound distracting and disruptive and would quickly leave websites that forced sound on them. If sound doesn't serve a purpose, it's often better if it is not there at all.

- Will it be better with sound or without?
- Are users more likely to use it if it has sound?
- What purposes does it serve for the users? What problem(s) is sound solving?
- What purpose should sound serve in our product—what functions will it play and why?
- Do we want consumers to associate any sounds with our product or brand, and if so, which sounds, and why?
- How do we want sound to impact users' overall impressions and experience of the product or brand?
- Do we want to evoke any associations in terms of adjectives with our product, and if so, which sounds and adjectives, and why?
- Do we want to evoke specific feelings or emotions, and if so, which sounds and feelings and why?
- Who are our users, and who else in their environment may be impacted by any sounds we use?
- Do we want to increase user engagement and interest in our brand with the help of product sounds, and if so, how?
- If sound is only there for aesthetic reasons, will it add enough to the product to justify its inclusion? (See Yanagisawa et al., 2009.)

5.1.2 How Can Sound Facilitate Basic User Actions?

A companion to the questions above are questions about what the sound is adding to users' experience. There are many times a product will function better without sound. I recently went to a self-checkout at a store and had scanned the first item nine times before I realized there was no beep to let me know it had been scanned: I had to wait for a clerk to reset the machine, and I explained that "I was expecting it to beep to let me know

I had scanned it." She said, "Yeah, it's a problem: that's what a lot of people say." Knowing the expectations of your audience can be important to understanding how sound should function.

- What do we want users to know?
- What do we want to draw their attention to?
- Is the sound functional?
- Does sound improve the navigation and user interface experience?
- Does sound improve another aspect of usability? Or does sound improve users' brand experience?
- What does the sound *need* to do?
- What do users think our sound is for, or should be for? What functions should sound serve in the product?
- Do we want to try to impact actual or perceived product performance with our product sounds, and if so, which sounds and why?
- In what user situations and at what times of day will the product sounds be experienced?

5.1.3 What Customer Touch-Points Can We Use Sound With?

We may have multiple points of contact with our customers. Thinking through how each of these touch-points might take advantage of sound, and asking users where they may have heard, or want to hear sound, may be useful to the product. For instance, if we were designing on behalf of a car company, we may design sounds for the car; for television, web, and radio advertising; for the website; for a mobile app; for the dealerships or any other associated retail experiences; and for other areas. There are multiple places where we have an opportunity to use sound to our advantage.

- Where and when should we have sound?
- Where can we leverage sound for a complete sensory experience of the product?
- Can sound extend beyond the product and become a part of the overall brand?

  And, of course, interactions with the product itself can also make sounds.

- Are there any sounds associated with physical interactions with the product (e.g., button clicks, switches)?

- Are these sounds pleasing, or are there ways that the interactions can be engineered to be more pleasing?
- How would users describe the sound produced when interacting with the product (e.g., button clicks, switches, sliding motions)?
- Does the sound align with user expectations of how the interaction should sound?

Don't forget that packaging, too, makes a sound.

- Does the packaging of the product produce any notable sounds during unboxing or handling, and what purpose might it have if we design packaging for a complete sensory experience?

5.1.4 What Sounds Should Be Included or Not Included, and Why? Where and When Should Sound Be Used?

This question is also closely related to the above questions. We might ask users where they feel a product needs sound, or we might test the use of sound in different circumstances. We may ask users, for instance, "Where do you think someone should or would use sound?" and see if they believe sound should be used in situations we haven't even thought of.

- Why is the use of sound better than the alternatives (e.g., haptic or visual feedback)?
- Under what circumstances would users want an alert within this product? (Do we need to warn users about anything?)

The addition of sound can be helpful if it reinforces certain aspects of information already provided in another format, because in different contexts we may want additional or alternate modalities. If I'm driving in a busy city, I want my GPS to tell me where I must turn with sound, not with visuals, so that I can keep my eyes on where I am going. On the other hand, if I'm walking, I'm embarrassed by the voice and just want the visuals. Or if I'm visually impaired, the addition of sound might help me to understand something important that I might miss on a small mobile screen, for instance.

5.1.5 What Are My Competitors Doing?

- Is there a company outside our space with a similar positioning as us, and what are they doing with sound?

- How do users feel about their sound? Are users annoyed by their sound, and if so, is it because they don't like the specific sounds chosen, or simply because there is sound?
- How are my competitors allowing for control over their sound?
- What do users remember about the sound in a current product they use?
- Do users recognize the sound?
- What product sounds from this type of product are ordinarily experienced or noticed in the widest sense?
- Can they name the brand or product the sound is from?

If we are developing a product for which there are many other examples that have been in existence for some time, what are the common conventions or sound tropes that competitors use? And do we want to stick with convention, or do we want to risk deviating from the convention in order to stand out? Are there any competitor sounds generated by the product that users find enjoyable or distracting?

As you proceed to designing sounds, think about asking the following questions.

- Do users think the sound enhances or detracts from the brand's identity?
- What impression does the sound leave on users about the brand?
- Is our sound preferable to that of the competition?
- If users have used similar products with different interaction sounds, how do the sounds of this product compare?
- Are there any specific interaction sounds from other products that users prefer over the sounds of this product?
- Does the sound stand out from competitors' sounds or sonic identities?
- What makes this sound unique or different from others users have heard?

5.1.6 What Are Our Users' Sound Preferences?

If we have a specific user base, we may also wish to query users about their preferences when it comes to sound.

- Do they prefer a specific type of sound (e.g., melodic, percussive, minimalist)?
- Are there any particular sounds that they find pleasing or displeasing?
- Do they have any personal preferences or biases toward certain types of sounds?

As described earlier, we lose higher frequencies as we age, so a younger demographic may have a different response to, or preference for, certain types of sounds in comparison to an older demographic, who may find "tinkly" sounds hard to hear. Think clearly about the user base for the products.

- What types of music do users listen to, and how might we leverage aspects of it for our own brand or product?
- Are there any sounds associated with physical interactions with the product (e.g., button clicks, switches)?
- Are these sounds pleasing, or are there ways that the interactions can be engineered to be more pleasing for our particular user base?

5.1.7 What Options or Features Would Users Like to See Related to the Sounds? What Controls over Sound Should the User Have? Would Users Like the Ability to Customize or Adjust the Sounds?

It's good to find out from our users what options they may want to have for sound. For some products, users may want to have the opportunity to customize sounds, or the option to turn off sounds. For other products, these types of settings may reduce our brand's impact if we allow for customization. It may be impractical (or even illegal, in safety cases) to allow for customization, or it may just not be necessary. Find out what users want when it comes to sound.

5.1.8 How Do We Want Our Users to Feel?

Although this question also falls under basic user experience research, rather than being specific to sound, it's useful to think of this question in the context of discussions with the design team about sound, because sound plays such a strong role in driving emotional response. Answering this question early in the design process can influence the ways in which designers think about sound. If we have a product that we really want users to connect with and feel attached to, then answering this question will also help to answer other questions about whether the product should have sound, and when and where to use sound. For example, let's say we have a mobile habit tracking application that is meant to encourage users to open it every day and track progress on a habit. We want them to feel encouraged and proud. So, knowing our goals with the product from the start tells us that including reward and positive feedback sounds would be beneficial. We can then take a look at not just competitor apps, but other devices that give positive feedback or auditory rewards (such as slot machines or video games), and explore how those are using sound. Should we include negative sounds to discourage particular behaviors? And how might those influence our users?

5.2 Brand and Product Identity

We should understand how sounds meet or defy our user expectations in the context of our own brand identity (assuming we have one). When thinking about sound, or designing and testing sound, we should be assessing if sounds are consistent with the brand's voice and identity. Is sound going to make our brand more memorable? Will sound help to define us (or our users—think Mac start-up sounds) or differentiate us from our competitors? What is sound adding to our overall brand?

5.2.1 What Existing Sounds Are Associated with Our Brand (If Any)?

- Which of our existing sounds is the most memorable?
- Which of the sounds best represents our values?
- Do users remember our existing sounds, and what do they think of them?

5.2.2 What Does or Could Sound Bring to the Brand?

- Is sound going to make our brand more memorable (or does it already)?
- Does sound help to make an emotional connection with our users?
- Does it help to define our users or differentiate us from competitors?
- What is sound adding to our overall brand?
- What impression does the sound leave on users about the brand?
- What is their first impression of the sound, versus their impression over a longer time frame (e.g., do they get tired of it quickly)?
- Do users think the sound enhances or detracts from the brand's identity?
- Do any of the sounds stand out or remain memorable to users after using the product?
- Are there any specific sounds that users can recall easily even after some time has passed?

5.2.3 Is the Sound Cohesive with the Product and/or Company's Brand?

- Is the sound cohesive with the rest of the audio/visual design of the product?
- Do the sounds create a consistent and coherent experience throughout the product?
- Are the sounds similar in style and tone across different actions or events?
- Do these sounds reflect the product's and/or brand's personality or intended user experience?
- Do the sounds match the context or nature of the product?
- If the brand has visual elements like logos or colors, do users feel the sound complements them well? A bright, happy yellow logo is unlikely to pair well with a low-frequency, distorted sound, for instance.

Think through multimodal interactions, and ways in which we can leverage cross-modal interactions.

- What are users' overall thoughts on the sounds?
- Do they feel that the sounds are appropriate for the context or purpose of the product?
- Are there any specific sounds associated with the product that stand out to them?

5.3 Sensory Impact

When asking questions specific to product interaction sounds, you can dive deeper into users' experiences and perceptions regarding the sounds associated with interactions with the product. These are questions about how sound engages different senses and contributes to the overall sensory experience.

5.3.1 Do Users Like the Sound? How Does Sound Make Users Feel?

- Does sound increase engagement and/or enjoyment/satisfaction?
- Does sound improve the likability of the product?
- What are users' first impressions of the sound?
- How would users describe the sound when interacting with the product (e.g., button clicks, switches, sliding motions)?

- Does the sound align with users' expectations of how the interaction should sound?
- What emotions do users associate with the sound?
- How pleasing is the sound?
- How stimulating is the sound?
- Does this product make users feel good?
- Is it fun?
- Does it empower the users?
- Which words capture the interaction/sound?
- Does the sound make users feel more motivated?
- Does the product sound powerful?
- Do the sounds feel responsive and satisfying?
- Do these feelings change over time the more users hear the sound?
- Do these sounds contribute to users' perceptions of the product's quality or value?
- Are there any specific aspects of the sounds that users find particularly pleasing or displeasing?
- Do users notice the sound or notice a difference in the sound?

5.3.2 What Is the Sound Communicating to Users?

- What does the sound connote (i.e., what associations do users have with the sound)?
- What mental imagery is evoked?
- Do the sounds remind users of anything else?
- Do users find it easy to understand or interpret the meaning behind the sounds?
- Does the sound remind users of anything else they've heard?
- Does the sound evoke any memories of using this kind of product?

5.3.3 Is Our Sound Better than That of the Competition? If So How/Why?

- If our users have used similar products, how do the sounds of this product compare?

- Are there any specific interaction sounds from other products that users prefer over the sounds of this product?
- Does the sound stand out from competitors' sounds or sonic identities? What makes this sound unique or different from others they've heard?

5.3.4 Which Configuration Is the Least Annoying or Most Preferred?

- Which configuration do users prefer?
- Which properties are most relevant for users?
- Which configuration has the highest aesthetic value?
- Can users differentiate between different versions?
- Do users feel that there is enough variety in the sounds to keep the interactions engaging and interesting?

5.4 Cognitive Impact, Usability, and Interaction Quality

Particularly in the case of interface sounds and usability, it is useful to explore how sound impacts memory retention and recall of information or features. In complex interactions, it may be necessary to investigate whether sounds simplify or complicate the cognitive process. Explore whether sounds intuitively guide the user, aiding in navigation and understanding, and assess the efficiency of sound as a feedback mechanism for user actions. Do the sounds guide users in correcting errors or navigating complex tasks?

5.4.1 Which Properties of the Sounds Are Most Relevant to Users?

- Which properties or attributes of the sound are users focusing on?
- Do different individuals focus on different attributes or aspects of the sound? Why or why not?

5.4.2 How Does the Sound Affect Users' Ability to Learn About and Adapt to the Interface or Product?

- Are the sounds distinctive enough to differentiate between different actions or states of the product?
- Do the sounds guide users in correcting errors or navigating complex tasks?

- Is the sound too distracting?
- Does task completion time improve with sound? Does task enjoyment increase with sound?
- Does sound improve any learning curve associated with the product?
- Does sound improve productivity?
- Are there metrics that we can track regarding the inclusion of sound, such as improved task performance times, improved navigation of menu items in an interface, decreased error rates, increased recall of information, better response times to alerts, or some other metric?

5.4.3 Is the Sound Informative? What Information Does It Give Our Users?

- What function do users associate with sounds?
- Which configuration has the highest functional value?
- Do the sounds help users understand the actions or events occurring in the product?
- Are the sounds relevant to the actions or events they represent?
- Was the interface easier to understand with sound (compared with no sound)?
- Do the sounds distract users from an intended task or goal?
- Do users find the notification sounds informative and easy to understand?
- Do the sounds contribute to the intuitive use of the product, making it easier to navigate or interact with?
- Do users receive clear auditory feedback when performing actions or interacting with the product?
- How does *perceived* effort vary with and without sound? (Raw metrics may not change, but if the perception of improvement is made with sound, is that enough for your product?)
- Which configuration has the highest functional value?

5.4.4 Do Users Want Control over Sound Settings and/or Customizability?

- How should users control the sound if we give them options? This might include volume, customization, or automatically adjusting according to the background noise level.

- Do any sounds feel too intense or jarring?
- Are there any new sounds that were felt to be missing, or variations users would like to see implemented?
- Are there any specific changes or modifications users would recommend to enhance the overall sound experience?

5.4.5 Are the Alert Sounds Clear and Distinguishable from Other Sounds in the User Environment?

If our users are going to be using our product in a busy environment, will any important alerts be masked by other sounds in that environment, or will it be confusing which product is making an alert if there is the potential for multiple other alerts? For instance, imagine a busy kitchen—are the beeps warning that the food is ready in the microwave the same frequency as the sounds in a bunch of stoves on the market? If it is, then users will waste time figuring out which device is trying to alert them. How can we make that easier for our users?

5.4.6 Does Sound Improve Discoverability?

Does the sound draw our users' attention to what we intend it to? We may want to use progressive disclosure through sound to help users learn a complex interface. Our health tracking device might start with some basic feedback on heart rate monitoring and provide positive rewards for activity, but as users progress and become comfortable with the device, the sound might alert them to dig deeper into the options to track heart rate variability over time, for instance, and to stop providing the user with sound for simple heart rate in favor of other metrics.

5.4.7 What Parameters About Repetition Might We Need to Be Aware Of?

Will sound play once and stop, or repeat? Or play a few times then stop? My car's seat belt warning sound will play as long as I'm not buckled. Is that necessary because of a regulation, or has my car manufacturer decided that it should just beep until the seat belt is buckled? A few beeps would be more than enough to warn me that the belt is unbuckled: any more, and it's unnecessary and annoying.[1] How a sound plays back can impact the reception of the sound, so it's worth thinking not just *what* sounds you use, but how, when, and where they play back.

5.5 Accessibility and Inclusivity

As discussed previously, hearing is a sense that noticeably deteriorates with age, and an alternative should always (where possible) be provided for users with hearing impairments. Likewise, if the product is used by those with visual impairments, making sure that they understand the sonic interactions is key to making your product accessible to a wide audience. Customization of sounds can help users who may have difficulties with certain frequencies—for instance, you may offer the ability to drop an entire set of interface sounds up or down an octave to make them easier to hear for people who have hearing loss in a particular range. Volume controls can let users set the volume to a level that is comfortable for them.

5.5.1 Does Sound Make the Product More Accessible to Users with Different Needs?

- What accessibility options do we need for those with hearing impairments? For users who are visually impaired?
- Does sound improve accessibility?
- Are there alternative modalities (images, haptics) for information that is presented with sound?
- Sound may provide functional use of our product by users who are visually impaired, but if the information is not given in another modality, it will be inaccessible to users who are hearing impaired.
- Is there an opportunity to allow for customization of information output for different users?
- Can users who are visually or hearing impaired use our product without any more difficulty than someone without impairments?

5.5.2 Are All Sounds Appropriate and Sensitive to Different Cultures?

Far from being the universal language promised by countless adages, music, and to a lesser extent sound preference, has different valences in different cultures, to some extent. One study showed that Western concepts of major keys being positive and minor keys being negative is opposite for northwest Pakistani listeners (Lahdelma et al., 2021). Another found differences between American and South Korean listener preferences for auditory icons (Jeon et al., 2015). Musical preferences are strongly linked to social class

(Cian et al., 2022). Age, gender, and education level also influence sound preference (e.g., people tend to prefer natural sounds as they age; Yu and Kang, 2010). Cultural and regional differences can complicate the perceptual evaluation of sound (Kim et al., 2012). These differences can depend on the product, however, and so testing should be conducted where relevant (Romagny et al., 2024).

5.6 Contextual and Trend Analysis Questions

It is also necessary to consider the conditions under which someone might use your product and expose not just the user but also those around them to the sound. Moreover, it might be relevant to think about the user's situation or state under which they are using the product and how sound might cause additional stress or help the user to feel relaxed (e.g., busy, relaxed, in public spaces). In exploring robot sounds, for instance, we might consider the fact that a carebot at home would need to be soothing, and could speak softly in an intimate way; however, a robot that is a service robot designed to direct people in a shopping mall would not need to have those qualities, unless perhaps interacting with a child.

A related consideration is the ambient noise levels we might expect the product to be used in. A product designed to be used on a factory floor may need to be much louder and have higher frequency sounds than a product designed for the home. Is there a way the designer should consider adapting the sound to the loudness of the environment, using a built-in microphone? While we have adaptive brightness on our phones, adaptive sound, although built into some phones, is not a common feature. I don't mind the sound of my MacBook booting up in public, since the sound is quite brief, but if it were louder, or longer, or happens in some quiet locations like a library, that sound would bother me (and others around me). If such adaptations might be useful, can those be built into the product?

5.6.1 When Would Users Turn Off the Sound, and Under What Circumstances Is It Appropriate to Have Sound?

- Will the product be used by a variety of types of audiences, and what influence should the intended audience (if it changes) have on the sounds?

- Should sounds adapt to my audience? For example, if our product is mostly used by adults but may also be used by small children, we may want sounds that can adapt to the user to help them, or make them feel comfortable, and these sounds may be different for different user groups. With the technology we are using, is it possible to make such adaptations?
- What about the mood of the user? If I'm really frustrated trying to figure something out, sounds that are repeatedly telling me I'm making a mistake may push me over the deep end in terms of my frustration levels—should we consider the number of times an error tone plays, or adapt to user frustration by, for instance, reducing error notifications?
- What adaptations might the product have for users' states of mind?
- How often will users hear the sound in context?

It's always worth finding out if your users want sound. As seen in the PlayStation example, some users like the boot-up sound—it's part of their experience and gears them (and their pets) up to play. Other users don't want sound at all. Most just want the opportunity to customize the sound or the option to turn it off. If you can't include the option to turn off sound, are you sure that your users want sound in all contexts?

5.6.2 Does Our Sound Design Align with Current User Experience and Technology Trends? Do We Want It To?

As with visual design, although I would argue to a lesser extent, sound goes through trends. What are the current trends in terms of user experience of products, or technologies that may be similar?

5.6.3 What Are Our Constraints?

- Are there regulatory issues we need to be aware of when it comes to the product sound?
- What are the technological options for offering users customizability, volume adjustment, and other features?
- Will the product have a high-quality speaker?
- Is it something like software, which may play back on a variety of different devices with different quality speakers?
- What are our technological constraints in terms of hardware or memory?

You can spend all the money you want creating the perfect sound for your product, but if it sounds terrible coming out of the ten-cent speaker the engineers put on the device, it's not going to be well received. Have you tried to play back the sound on different devices (if applicable), in different locations, or in different configurations of your product, and will that impact your user?

- Are the sounds loud enough or soft enough?
- How loud should the sound be?
- Would users like the ability to customize or adjust the sounds?
- Do these customizations include changing the sound, or just changing the volume?

5.6.4 How Much Repetition Is There Going to Be?

- Might a user response to a sound change over time?
- Will sound play once and stop, or repeat? Or play a few times and then stop?

Repetition has been discussed a few times, but understanding how often users may take a particular action that causes a sound is relevant to the design of that sound. As discussed earlier, sounds can be pleasing the first few times we hear them, but over time may come to irritate us. Thus, knowing how often a sound is being played in common use cases can be very useful, and thinking of ways to adapt a sound, provide variety, or make other options available may be helpful. Perhaps instead of exactly the same sound being used, we can add some variability into the sound (as is common in video games). For example, the sound may be played from a folder of optional related sounds or generated in real time with minor variations.

5.7 General User Experience Sound-Related Questions

If you have a specific user base, it may be worth thinking about asking them some of the following types of questions in any surveys or questionnaires being conducted.

- Do you prefer a specific type of sound (e.g., melodic, percussive, minimalist)?

- Are there any particular sounds that you find pleasing or displeasing?
- Do you have any personal preferences or biases toward certain types of sounds?
- Do you feel the sound complements the brand/product's visuals well?
- Do you recognize the sound? Can you name the brand or product the sound is from?
- Which of our sounds is most memorable?
- Which of the sounds best represents our values?
- What function do you associate with this sound?
- Can you differentiate between configurations?
- What properties are most relevant to you?
- How pleasing is the sound of this product to you?
- How stimulating is the sound of this product to you?
- Does this product sound powerful?
- Does the sound evoke any memories of using this kind of product?
- Does this product make you feel good? Is it fun? Does it empower you?
- Can you control the sound?
- Do you prefer it with sound?
- Does sound make things quicker or easier?
- Is the sound too repetitive?
- Does the sound draw your attention to something?
- Do any sounds feel too intense or jarring?
- Are there any specific sounds associated with the product that stand out to you?
- Do the sounds feel responsive and satisfying?
- Which words capture your experience of the sound?
- Did you notice different configurations or versions of the sounds we presented? Which configuration is least annoying or most preferred? Which configuration has the highest aesthetic and functional value?
- What emotions does the sound evoke for you? What memories does it evoke in you?
- Is the sound in line with your expectations?

5.8 Conclusions

There are so many questions specifically relating to sound that can be incorporated into UX research at many stages of product design. Deciding which questions you need to ask, and when, is the first step in the research journey that will ensure effective user experiences. The UX research process should be ongoing throughout the design process. Asking questions about the sound is an iterative process that benefits from continuous questioning and feedback. Where possible, bring team members from different disciplines (e.g., design, development, marketing) into the question development process, in order to gain their diverse perspectives and specific needs.

The following chapters cover the methodologies that you can employ to answer some of these questions. There are no hard and fast rules about which methods map to which questions; however, some suggested methods are presented in the following chapters.

6 Testing Methods for the Subjective Measurement of Sound

Now that you have your research questions formulated, this chapter aims to provide an overview of the subjective methods and tools used to assess sonic user experiences. By *subjective*, I mean responses by users based on their own personal feelings or opinions, rather than through external validation or evidence (the subject of the following chapter). These methods are designed to gauge individual reactions and perceptions, which are inherently subjective. There are both qualitative and quantitative research methods for the subjective measurement of sound, and each of these methods has its own strengths and weaknesses.

When it comes to selecting a method, consider the timing of the test (where in the product life cycle the design is), the resources (including how many people are available as participants and researchers, the facilities available, the resources for analysis, and your own expertise and experience levels), the participants (who is needed, how many and what type of users, and how to recruit them), and the milieu the test is taking place in (i.e., what do different stakeholders want from a user test?)

Generally speaking, a mixed-methods approach (i.e., using more than one method to answer the questions) will be the most fruitful, since we can get the depth of experience captured by qualitative responses, and combine those with statistical data of a quantitative analysis that stakeholders tend to love. Since no subjective method is entirely free from bias, employing multiple methods and comparing their results can also enhance validity. When you are conducting qualitative research, consider adding some quantitative testing as well, because engineers and business folks understand the value of numbers and data, and are less likely to appreciate the value of qualitative work.

Subjective and self-reported measures rely on the idea that participants accurately can report their felt emotions, feelings, and experiences. This is not always the case, of course. Most of us have experienced times where we're not sure what we're feeling, let alone how to articulate that. Emotions are complex, and we can experience conflicting feelings at any time, as well.

In addition to issues with a user's ability to articulate their response, research has shown that survey format type (e.g., check all that apply or rating scales) influences results, as does the order of questions and the attributes selected in scales, the number of samples, and even the time of day of response (King et al., 2013). Pre-formed scales have undergone assessments of reliability, validity, and sensitivity. They can allow us some objectivity because they have been standardized, and replicability since they have been used widely.

There isn't room to cover every testing method in detail here. Moreover, I haven't covered general user experience testing methods that are not directly involved in measuring responses to sound (such as personas or user profiles). This lack of inclusion is not because they are not useful methods, but because I'm specifically focusing here on those methods that are applicable to testing the sound-related questions discussed in chapter 5. Further, there are methods that I have included that are drawn from sensory analysis and psychology that are not typically applied to user experience testing, but I believe they are relevant and useful with sound-related questions.

6.1 Observational Methods, Interviews, and Co-design

A variety of observational approaches exist that can be useful in user experience testing, including contextual inquiry (observing users using the product in real-world scenarios), cognitive walkthroughs, focus groups, think-aloud (Wang et al., 2022), diary studies, and other approaches. When it comes to sound, there are some minor considerations with these methods, particularly when it comes to talking aloud while a sound is playing, potentially obscuring other sounds. Observational methods can be very useful in the formative stage of the design process, to be developed through further iteration and testing.

Rich use scenarios, condensed descriptions of the uses of an application, are commonplace in user experience research. By visualizing how a user might interact with a system, designers can make informed decisions that align with real-world use. Detailed scenarios can highlight potential issues or areas of friction in the user journey. They help in communicating the vision and purpose of a design to stakeholders, ensuring everyone is on the same page. Scenarios can also be used in testing to guide participants through specific tasks or situations. An alternate approach used for sound design is the radio play method. This method is a rich use scenario presented in the form of a radio play (an audio dramatization). In this way, sounds can be incorporated directly into the scenario, beyond mere narration (Pirhonen et al., 2007).

Example: A team designing notification sounds for a smartwatch observed how nurses in an ICU responded to different alerts while multitasking. Researchers could develop rich use scenarios showing how overlapping alarms caused stress, and how subtle auditory differences could improve response prioritization. They could annotate the sequences with quotes from nurses to show how alert tones clashed or helped.

Co-design practice, closely related to *participatory design*, is a practice where stakeholders (including users) collaborate, co-creating during the design process. Co-design practices in sound have typically involved people who are visually, physically, or cognitively impaired (Cuadra et al., 2023), or are used in musical instrument design, due to the difficulties inherent in prototyping sound, although this area could be developed. The downsides are that typically it's only feasible to use such practices with very small groups of people. Nevertheless, the methods may include letting users create and test their own prototypes in their own homes or offices.

Example: In a co-design workshop for a fitness app, users create their own sound cues using voice, mouth sounds, and toy instruments to communicate motivation, fatigue, and encouragement. Designers gather and map these to potential app interactions, for future testing. For instance, a participant may say something like, "I liked the soft swoosh for a finished goal. It feels like an exhale. The bell felt too aggressive." The comments and sounds are then developed further and tested using other methods.

6.1.1 Interviews

Interviews are a common method to gather information about user needs, motivations, pain points, and behaviors. Interviews can be strictly structured, semi-structured, unstructured, or open. A strictly structured approach is similar to a questionnaire, in that all questions are predetermined and asked in a specific order. At the other end of the scale, an open interview is more like a guided conversation. Typically, we want at least some structure, that is, some specific questions that we ask every person we interview. A formal structured approach might involve a protocol for note-taking, and even a template of standard questions and how answers will be coded. Semi-structured interviews are perhaps most commonly used, and have questions or general themes decided in advance, but these questions can be followed up on, often in what is called laddering, where the answers are examined as much as possible to ascertain the *why* behind the answers. This approach allows the researcher to delve deeply into unexpected areas. Usually, naïve participants are used for interviews, and are asked to describe their experiences with a product, before, during, and/or after using a product.

Some example questions are the following.

- Can you describe the sound(s) you hear when using this product?
- How would you describe the volume of the product? Is it too loud, too soft, or just right?
- Are there any specific sounds or tones that stand out to you? If so, can you describe them?
- How would you compare the sound of this product to similar products you've used?
- How does the sound of the product make you feel (e.g., relaxed, annoyed, satisfied, disturbed)?
- Are there any sounds from the product that you find particularly pleasant or unpleasant?
- Does the sound of this product remind you of anything else?
- How important is the sound of this product in your overall experience?
- Does the sound provide any functional feedback (e.g., indicating the product is on or off, or a task is complete)?
- Are there moments when you wish the product would provide an auditory cue or feedback? Are there times you wish there was no sound?
- In what environments or settings do you typically use this product? How does the sound fit or clash with that environment?
- Do you ever adjust the sound settings on this product? If so, how and why?
- Are there additional sound-related features or options you wish this product had?

- How do people around you react to the sound of this product?
- Are there any comments or feedback you've received from others regarding the product's sound?
- Have you used other products with similar functions? How would you compare their sounds with this one?
- Are there any specific products you can think of that have a particularly good or bad sound experience?

6.1.2 Diary Studies and Design Probes

Diaries can be collected from users when wanting to understand how they interact with a product in their homes, rather than having an observer there. For instance, a user could write down all of the product sounds that they hear in a day, and write down their impressions of the sounds. Or they may only write down comments specific to the product under investigation. Longer-term understandings of product sounds can be achieved this way, as the user interacts with a product over days or weeks: Initial excitement about a product can wear off, and the user may not like the sounds after hearing them a few dozen times. Such diaries do not need to be written, but could be a video or audio recording at the end of each day, text messages, or even drawings instead of text to capture feelings and impressions (Zorea and Kushi, 2023).

Although this method is cheap for the researcher in terms of time and effort, the user drop-off rate can be high, statements can be ambiguous, and it can be difficult to follow up on responses later in time when the user may not remember what they were thinking. Generally, diary studies are superficial unless participants have a high level of commitment.

Design probes were developed by Gaver, Dunne, and Pacenti in the late 1990s. Participants are handed a package (the *probe*) with various artifacts such as maps, postcards, cameras, and diaries, and given evocative tasks to record their feelings and interactions in an attempt to capture inspirational responses through creative rather than analytical questions. Sound probes have been created to allow people to explore everyday sounds, for instance, to develop meaningful sound-based interfaces (Jain et al., 2019).

Example: Participants are given a *sound diary* app, with a button that records when pressed. Over a week, users capture sound moments that feel intrusive or satisfying and annotate them with short notes or emojis. The diaries could be reviewed together in a workshop to uncover themes like *sonic comfort*, *annoyance*, or *attentional shifts*. A probe kit could include the sound diary smartphone app, and some prompts, such as "Record a sound that caught your attention unexpectedly," "Make up a sound to match how you felt at lunch today," and "Draw the shape of a sound you liked this morning." Participants use the kit over a few days to document their responses to sounds in their environment, and the data are then analyzed for each participant and collectively.

6.1.3 Examples of Work Using These Methods for Sound Design

Mario Owczarzak (2019) used diary studies as one method to explore in-situ responses to Facebook Messenger sounds. He found that response depended a lot on context and environment of the sound, and that sound was preferred in active conversations and in passive conversations when expecting information. Unsurprisingly, participants did not like sounds in quiet public spaces. Questions were provided in a structured format, asking the same repeated questions daily, increasing the number of questions as the week went on. Owczarzak found, however, that many participants did not undertake the daily prompt entries asked of them.

One approach to gathering diaries is Stephen Barrass's "EarBenders" (1996). EarBenders is a collection of stories about everyday activities involving listening to sounds. The method consisted of a database that had four stages: situation description (key features of situation, using a story), situation analysis (details task, information and data structure), example lookup (a list of examples ranked by similarity in structure with description), and design synthesis (the examples to inform guidelines about sonic interfaces). By organizing answers and collecting similar responses, patterns begin to form and can be examined.

Kim and Yun (2018) used a variety of methods, including interviews, to explore the impression of camera shutter sounds. The study learned that keywords from interviews can serve as a clue to represent the impression of product sound. They took interview data and classified the text into positive and negative groups. They then performed a frequency analysis of keywords in each group, looking for commonalities and dissimilarities (e.g., *friendly* versus *not friendly*), and quantified the number of times keyword groups were used.

Moesgaard et al. (2020) explored co-design to develop concepts to enhance passenger experiences in the Copenhagen metro. They undertook a variety of sound sketching and prototyping activities to explore customers' preferences. They carried out a thematic analysis of notes, recordings, and images, as a means of developing clusters of meanings and patterns that emerged, and invited participants to take part in a co-creation workshop. One said, "It's hard to put sound experiences into words. Especially when you don't have anything to listen to beforehand. But when you start listening to some different sketches or sounds it becomes much more tangible. It's easier to imagine what sounds would work and what sounds are annoying to listen to. . . . After the sound sketch, I also realised that it would be nice to have a dynamic soundscape that changed over time." The authors note that "sound sketching was seen to be an effective method in getting participants to verbalise and discuss different sounds and sound concepts. . . . Based on the concept of photo elicitation, we suggest further investigating a concept of audio elicitation, where designers or researchers use a variety of sounds as triggers during an interview and ask participants to articulate and reflect on the different sounds they encounter and the experiences they prompt."

6.2 Difference/Discrimination Testing

Discrimination testing methods aim to determine if there is a perceivable difference between two or more products or experiences. These tests are often used when reformulating a product (e.g., changing an ingredient) and wanting to know if the change is detectable, or comparing a product with that of a competitor to see if there's a noticeable difference. The question is not *how much* difference exists, but rather *whether* a difference can be detected—for example, can listeners perceive a difference between a highly compressed sound and a high-fidelity version? How does the quality of our version compare with the quality of the sounds of our competitors?

There are a number of discrimination test types that are available, far more than can be covered here. These methods often involve forced-choice tasks, which can produce quantifiable data. Each test has an associated chance performance level, that is, the expected proportion of correct responses if participants guessed. The sensitivity of each method varies; for example, paired comparisons are generally more sensitive to small differences than triangle tests, but triangle tests are more reliable statistically due to lower guessing probability.

When it comes to sound, there are additional difficulties with paired comparison tests in that sounds are temporal, and so cannot be presented simultaneously, which is common in many difference pairings. The participant must remember the sounds that are under comparison. As such, there is a memory element at play. Repeated playback of sounds (either automated with interval timings, or under user control) will help to mitigate this effect. My preference is to let users control the playback, especially with naïve listeners who may be evaluating sound for the first time.

There may also be order effects, such that a similarity rating may be slightly different for the same pair, depending on which sound is played first. There are several choices for how one might present sound pairs to address these potential complications. To address this, researchers can either present each pair twice in both possible orders and average the ratings, or randomize presentation order across participants. A summary of discrimination testing methods is found in table 6.1.

Table 6.1
Some Common Discrimination Testing Methods

Method	Name of Test	Test Samples	Task
Forced choice	Paired comparison	A, B	Choose sound with most of attribute
Forced choice	3-alternative forced choice (3-AFC)	A, A^1, B (where A^1 represents a variant of A)	Choose sound with most of attribute
Forced choice	n-AFC	A–A^{n-1}, B	Choose sound with most of attribute
Matching	Duo-trio	Reference, A	Match sound to the reference
Matching	ABX	Reference A, reference B	Match X to either A or B
Odd one out	Triangle test	A, A^1, B	Choose the most different sound
Same-different	A-not-A	A, B	Decide if test sound (B) is the same as A
Difference from control	Difference from control (DFC)	A, B	Rate how different B is from A on an attribute scale
Preference	Paired preference	A, B	Choose preferred sound
Similarity rating	Pairwise dissimilarity task	All pairs from a set	Rate dissimilarity on a scale
Proximity rating	Polarized sensory positioning (PSP)	All sounds rated on polar opposites	Rate each sound on opposing attribute scales

Unspecified difference tests ask participants to simply determine whether a difference exists, without highlighting any particular attribute (e.g., "Is there a difference between sound A and sound B?"). Unspecified difference tests are ideal for situations where researchers might not be sure what changes are perceptible after making modifications to a sound, or when they don't want to bias the panelists' perceptions by highlighting a specific attribute. For instance, we might use such a test to determine if users notice a degradation in quality of sound if we want to use a cheaper speaker in a product. Examples include A-not-A, triangle tests and duo-trio tests.

In an *A-not-A test*, for instance, panelists receive two or more sounds and must determine if they perceive them as the same or different ("Does the sound you hear sound the same as sound A, or different?"; see figure 6.1). Another test may ask the participant to choose the odd one out (in a duo-trio or in a triangle test; see figures 6.2 and 6.3). In a duo-trio test, a variation on the triangle test, panelists are also given three sounds: two identical, and one different. Panelists are played a control/reference sound first, followed by the other two sounds. Panelists are then asked to identify the sound that is *more similar to or matches* the control sample.

ABX testing, while very similar, was developed specifically for audio at Bell Labs. In this case, the participant focuses on detecting differences between a test sound (X) and two other sounds (A and B). ABX tests may

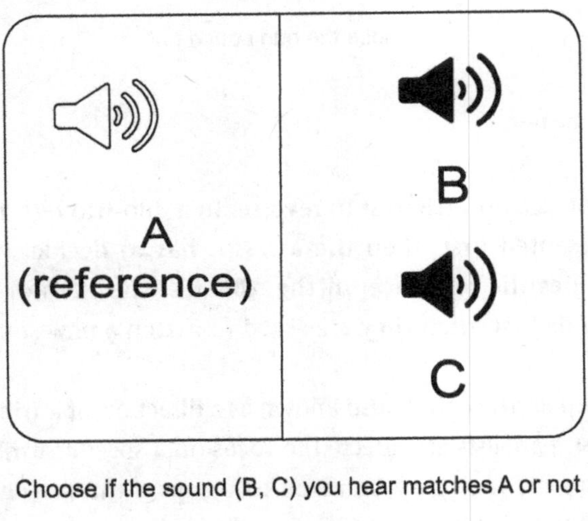

Figure 6.1
Example A-not-A test.

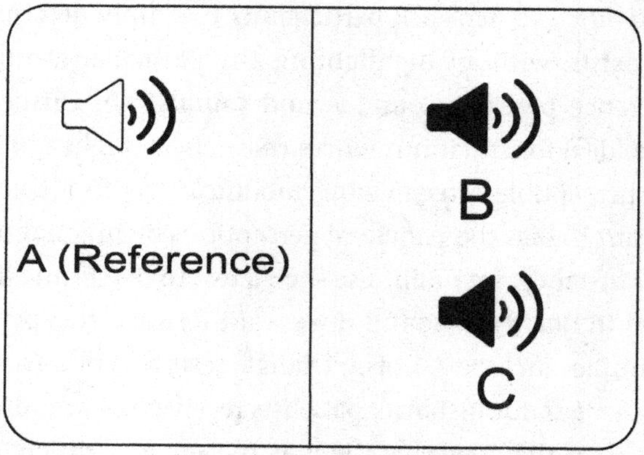

Choose the sound that matches the reference sound.

Figure 6.2
Example duo-trio test.

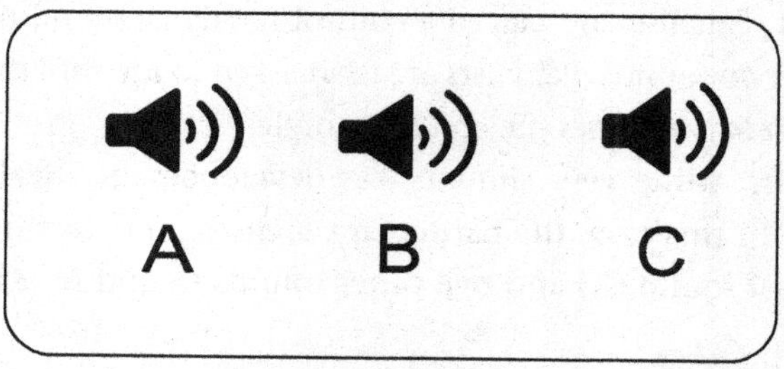

Choose the odd sound out.

Figure 6.3
Example triangle test.

be thought of as a duo-trio test in reverse: In a duo-trio test, the reference sound is presented first, then the assessor has to decide which of two sounds matches the reference; in the ABX test the assessor is given the two test sounds first, then they are asked to match a new sound to A or B (figure 6.4).

In a *specified difference* test, also known as a direction or attribute-specified difference test, panelists are directed to focus on a specific attribute to compare. For instance, they may be asked to compare the perceived loudness ("Which of these sounds is louder?"). Specified difference tests are often used when product developers are making a deliberate change in a particular attribute of a product and want to know if that specific change is detectable.

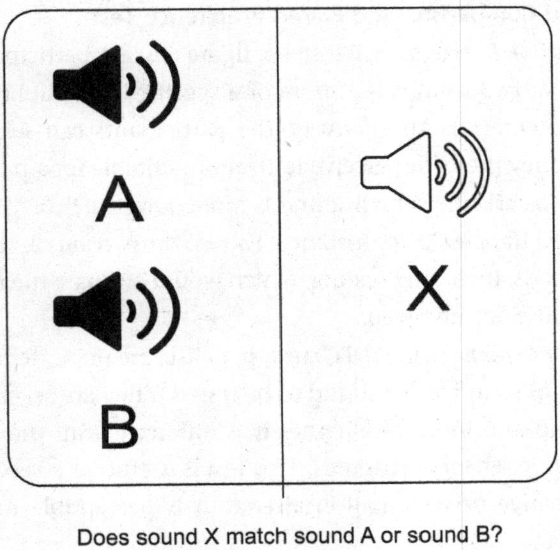

Figure 6.4
Example ABX test, an inverse of a duo-trio test.

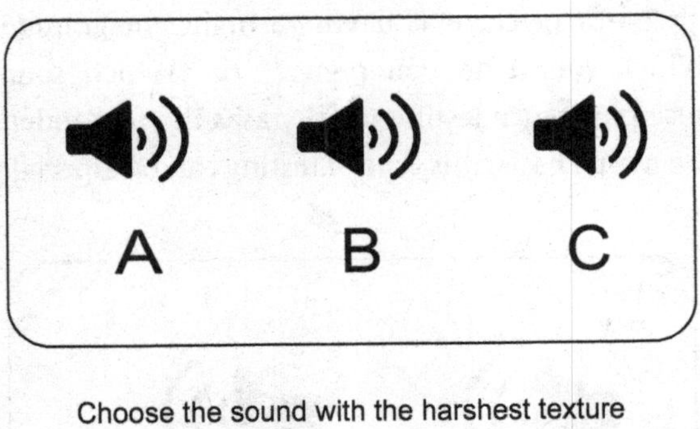

Figure 6.5
Example 3-AFC test.

For instance, participants may be asked to choose a sound with more of a specific attribute between two sounds (such as more timbral texture), in another discrimination test known as two-alternative forced choice (2-AFC), or a sound with the most of a specific attribute out of three samples (3-AFC; see figure 6.5). The n-alterative forced choice (n-AFC) test involves a participant choosing a sample from a group of two or more sounds that have more or less of a particular attribute (*n* refers to the number of choices, such as 2-AFC or 4-AFC).

6.2.1 Paired Comparison and Paired-Preference Test

Paired comparison tests (e.g., A/B test; see figure 6.6) ask participants to determine which of two sounds has more of a specified attribute (e.g., "Which sound is crunchier?"). Alternatively, the participants can select which of the pairings they prefer or perceive to be more suitable for a purpose. A participant may be asked, "Which sound is more powerful?" or "Which sound makes you feel happier?," for instance. Paired comparison can be done with various methods, including scaling, which would evaluate the relative magnitude of preference involved.

In a *difference from control (DFC)* test, panelists are presented with a reference or control sound and a sound to be tested. They are asked to evaluate the test sound and indicate whether it is different from the reference in terms of specific sensory attributes. This test is useful for assessing whether a product change or formulation alteration is perceptible to consumers (figure 6.7).

A *paired preference test*, on the other hand, compares sounds in pairs of stimuli, and asks the participant to determine which of the two sounds they prefer (see figure 6.8). For each pair, respondents could indicate which sound they prefer or perceive as having a higher magnitude of a certain attribute ("Which sound do you prefer?" or "Which sound has more bass?"). Because preference testing usually asks the respondent to evaluate the product on a total basis, this type of testing can be especially susceptible

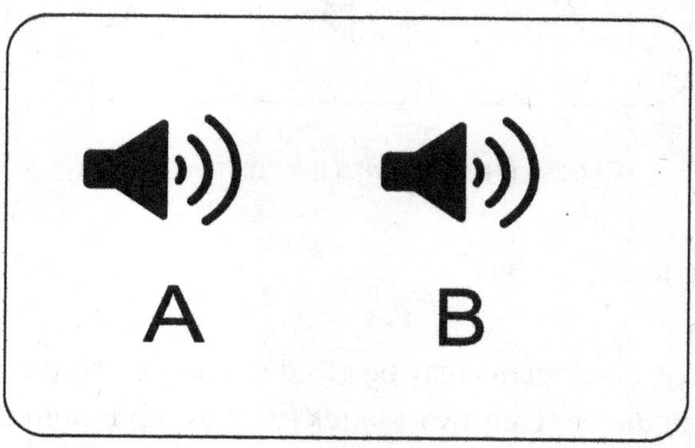

Figure 6.6
Example A/B comparison test.

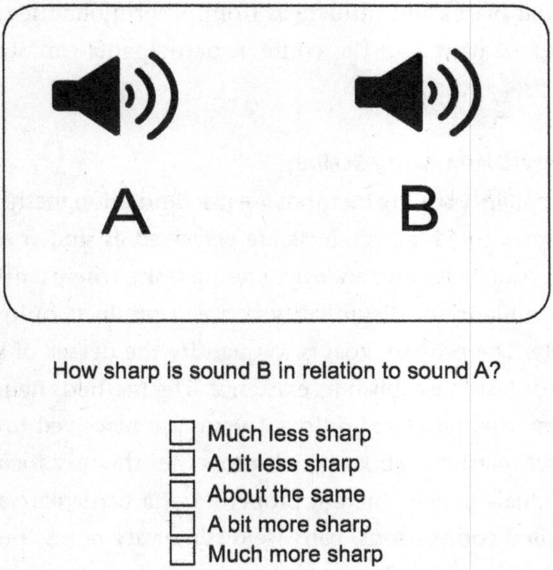

Figure 6.7
Example DFC test.

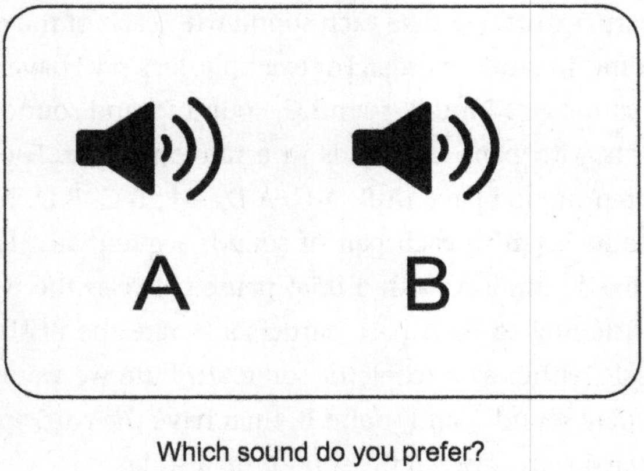

Figure 6.8
Example paired preference test.

to unintentional biases and influences from other modalities. Depending on the number of pairs and the context, participants can struggle to be consistent (Parizet et al., 2005).

6.2.2 Similarity/Dissimilarity Scaling

Similarity/dissimilarity scaling methods are discrimination methods designed to assess the *extent* to which products are perceived as similar or dissimilar. These methods usually involve scaling or rating tasks, where participants rate the perceived similarity or dissimilarity between products on a continuous or discrete scale. The primary goal is to quantify the degree of similarity or dissimilarity, not just to establish its existence. The methods help researchers understand how closely related different items are perceived to be, and are useful to answer questions about which properties the user focuses on, and whether individuals notice different properties of a particular sound. Methods include paired comparisons, pairwise dissimilarity rating, proximity ratings, and similarity scaling. The tests can be used on naïve assessors (thirty to sixty participants preferred) or expert assessors (at least fifteen preferred).

The *pairwise dissimilarity task (PDT)*, for instance, is a traditional method requiring participants to rate the perceived differences between pairs of stimuli. In PDT, participants compare each sound with each of the other sounds, one pair at a time, in random order. For example, let's say I have a set of alarm sounds (e.g., sound A, sound B, sound C, sound D, and sound E). I present the participants with pairs of sounds in a random order. For five sounds, there will be ten unique pairs (A-B, A-C, A-D, A-E, B-C, B-D, B-E, C-D, C-E, D-E). Participants listen to each pair of sounds sequentially. Each sound is played for a fixed duration, with a brief pause between the two sounds in a pair. After listening to each pair, participants rate the dissimilarity on a predefined scale, either as a whole or some attribute we want to measure. For example, play sound A and sound B, then have the participants rate the similarity in harshness between the sounds on a scale:

1 = Very similar

2 = Similar

3 = Moderately similar

4 = Moderately dissimilar

5 = Dissimilar

6 = Very dissimilar

Table 6.2
Example Dissimilarity Matrix, Numerical and Shaded Matrix

	A	B	C	D	E
A	0	4.2	2.5	5	3.8
B	4.1	0	3.7	4.8	2.1
C	2.5	3.6	0	4.5	3.2
D	5.4	4.9	4.4	0	5
E	3.6	2.2	3.3	5.1	0

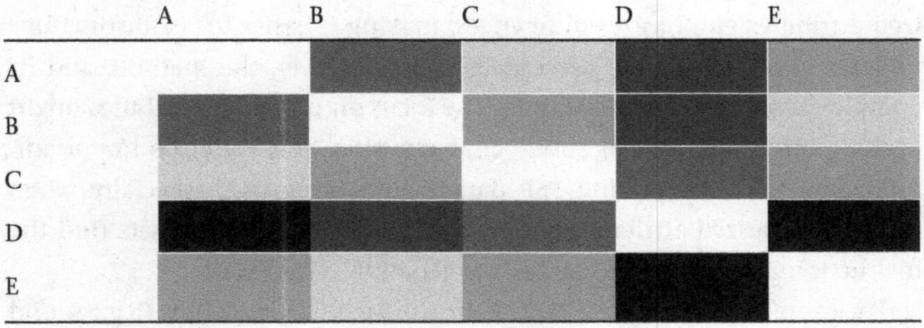

Record the dissimilarity ratings for each pair. Calculate the mean dissimilarity rating for each pair. Use multidimensional scaling (MDS) to visualize the perceptual space of the sounds based on the dissimilarity ratings. A dissimilarity matrix is created, to easily visually compare the results, and identify clusters or patterns in the perceptual differences between sounds. Table 6.2 is an example of a dissimilarity matrix, where each cell represents the mean dissimilarity rating given by participants for the corresponding pair of sounds. It's common to give the numbers a shaded value, where, for instance, 0 would be black and 6 would be white on the scale. This way, a quick glance can reveal patterns in the data.

Proximity rating is a type of paired comparison where participants provide a numerical assessment of dissimilarity between pairs of stimuli. For instance, we ask the participant to indicate whether the sounds are extremely similar or extremely dissimilar, and rate them on a scale. Polarized sensory positioning (PSP), is a common form of proximity rating.

Polarized sensory positioning is used to investigate the perceptual differences and similarities between sounds based on sensory attributes. The name *polarized* comes from the fact that this method aims to maximize the differences between sounds by emphasizing polar opposites in attributes (e.g., *dull* versus *bright*). Trained panelists first generate a list of these sensory attributes that describe the products. From the generated list, a subset of polarized attributes (pairs of opposites) is selected. Panelists rate each product on each polarized attribute using a scale. The ratings are then mapped to a multidimensional space where each product is positioned.

One advantage of the method is that comparing sounds to references is straightforward. It is ideal for distinguishing between very similar sounds, and does not require any universally agreed-upon vocabulary. Using polarized attributes emphasizes differences, making it easier to see distinctions between products on the perceptual map. However, the method, and its complex analysis, takes some time. The focus on polarized attributes might exclude other nuanced or subtle characteristics that could be important. Analyzing and interpreting PSP data require expertise, especially when multiple polarized attributes are involved. Moreover, many users find the instructions confusing, so detailed explanation is necessary.

For example, let's say we have four similar sounds to test, (e.g., sound A, sound B, sound C, sound D). We decide on a bipolar scale, for example, pleasant to unpleasant, melodious to harsh, or attention-grabbing to subtle. Usually these are Likert scales where, for instance, very unpleasant may be numbered 1 and very pleasant 7 (figure 6.9). A suitable question might be, "Rate the following sound on a scale of unpleasant to pleasant."

For each user, given four sounds (A, B, C, D), we might end up with a matrix like table 6.3. We do this for each participant. There is no limitation to the number of samples, although too many samples can become time consuming and tiring. Stable data usually require at least thirty participants.

Data analysis varies according to the specific type of PSP test and data, and can vary from multifactor analysis, principal component analysis, multidimensional scaling, or generalized Procrustes analysis. For this specific test, a principal component analysis is usually a good choice to visualize

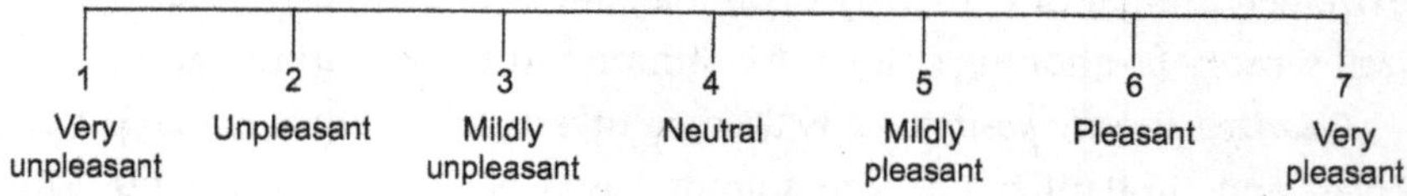

Figure 6.9
Example seven-point scale.

Table 6.3
Polarized Sensory Positioning Example Data Matrix Result

Sound	Unpleasant–Pleasant	Low–High	Short–Long	Harsh–Melodious	Subtle–Arresting
A	2	5	3	2	1
B	6	2	6	5	4
C	3	4	2	3	5
D	7	1	5	7	6

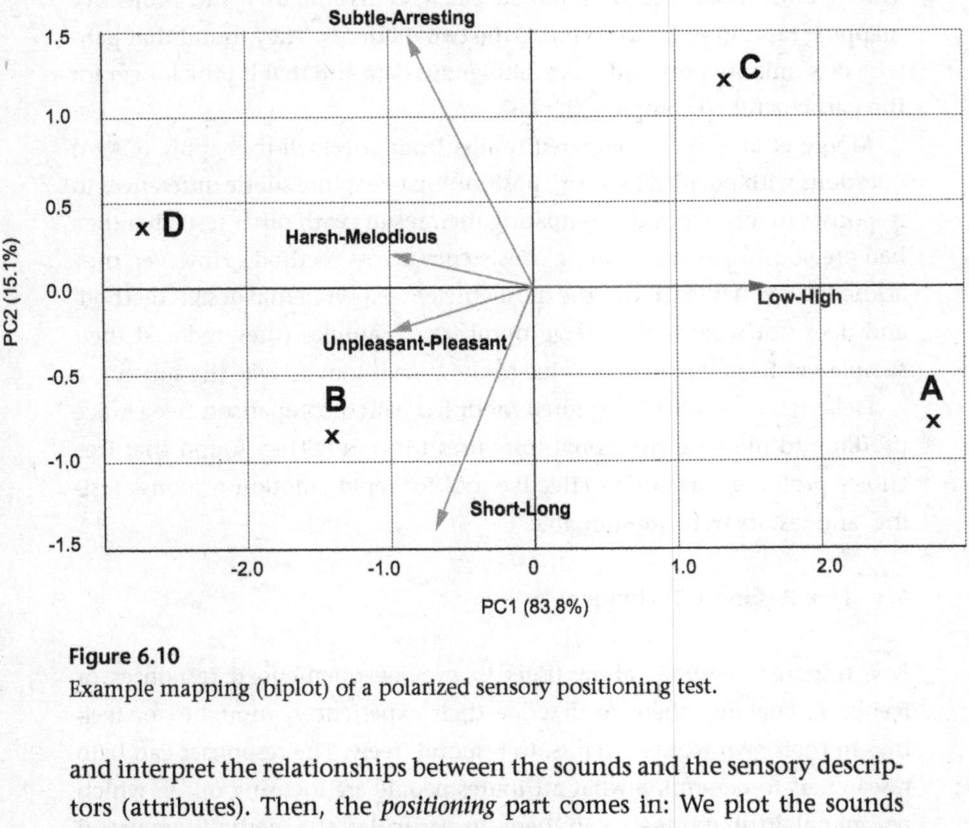

Figure 6.10
Example mapping (biplot) of a polarized sensory positioning test.

and interpret the relationships between the sounds and the sensory descriptors (attributes). Then, the *positioning* part comes in: We plot the sounds on a two-dimensional coordinate where each dimension represents one of the bipolar scales. For example, we might end up with our samples positioned as in figure 6.10. The four sounds (A, B, C, D) are plotted in the space defined by the first two principal components (PC1 and PC2). The arrows represent the original sensory attributes (like harsh–melodious or short–long). The direction and length of each arrow show how strongly that attribute contributes to each principal component. Sounds close together (e.g., A and C) are more similar in their sensory profiles. Arrows pointing in a similar direction are positively correlated (e.g., subtle–arresting and harsh–melodious). A sound that lies far along a vector (e.g., D along harsh–melodious) is strongly associated with that sensory attribute.

6.2.3 Examples of Discrimination and Difference Testing for Sound Design

Vowels and Mason (2020) explored pairwise dissimilarity and projective mapping tasks in order to compare the two methods. They found that pairwise dissimilarity provided more ambiguous data and that it took longer for the participants to complete the task.

Moore et al. (2019) compared results from check-all-that-apply (CATA) questions with polarized sensory positioning to explore subtle differences in responses to robot sounds, comparing their results with other tests that they had previously run, and finding consistency across methods. However, they acknowledge that PSP is not the most efficient experimental design method, and does not scale well to large numbers of samples (they reduced their sample size from twenty to six due to the time it took to take the survey).

Liebetrau et al. (2013) explored modified paired comparison free choice profiling to measure emotional responses to sound. They found that free choice profiling was not an effective tool for rapid emotion response testing, and results were questionable.

6.3 Free Response Techniques

Free response methods allow users to provide open-ended responses or feedback, enabling them to describe their experiences, thoughts, or feelings in their own words—that is, to respond freely. The responses can help researchers to determine what attributes people are focusing on, or which perceptual attributes matter to them. In particular, the methods are useful when asking questions such as "What does this sound mean to our users?" Not giving them prescribed responses from which to choose means we can capture all of the meanings we haven't even thought of, that may be specific to individuals.

Free verbalization is a method in which the user is presented a task or object, and then asked to spontaneously respond to that object, in their own words. Free verbalization has the benefit of telling us what sound characteristics users are focusing on—such as the semantic meaning or the acoustic properties—and what the sound connotes to users. Free verbalization has been used as a step to create the terminology for other types of testing, such as the attributes to be used for a semantic differential scale (Nosulenko and Samoylenko, 1997). Example questions include "Describe the sound you heard" and "What does this sound mean to you?" Further, free verbalization is the best method to determine what mental imagery is evoked, or what is connoted by the sounds, since it doesn't constrain the users to pre-determined terminology. Limiting the word count ("using one word, describe the sound . . .") can make analysis easier, but can miss some of the nuances that people may want to add to their descriptions.

Some examples of responses to some robot sounds I have explored, for instance, were:

- "Happy and excited!!"
- "Dancing, yay, accomplishment"
- "Light and friendly happy, sincerity"
- "Burst of electrical current"
- "Very sad Robot, Wall-E-esque"
- "Hello accomplice."

For longer sounds or music pieces, a longer paragraph can be instigated by prompting the listener to respond with a description of a scene. For instance, "Imagine this music was the soundtrack to a movie. Describe what's happening in the scene." I used this method in my PhD thesis exploring industrial music (Collins, 2003). The following were two different responses:

> Militant and futuristic like the future scenes of *The Terminator* in that AI run the world except clean like a shiny factory assembly line. These are machines making machines. In this version the AIs have already won against mankind and rule the future but are still constantly at war. It is the rhythm they live by. They are servants to themselves now. This is analogous of what they disliked in humans. They have pursued their vision so far they cannot stop, they emulate humans even though that is what they sought to destroy. It is all they know. They cannot change.

> A rebel force, underground gathering forces/equipment before an overland revolt. Machinery with cogs turning in a dark, dirty factory. An explosion to signify the beginning of the revolt, then a pause, quiet as they embark from the sewers. There is fighting, looting and ransacking as we see shocked "normal" citizens.

There are challenges with the free verbalization approach. It can be difficult to accurately verbalize abstract imagery in one's mind, and thus information is always incomplete. More notable is the challenge of analyzing the responses. Samoylenko et al. (1996) and Tagg (2013) designed systems of manually coding and categorizing terminology. Their methods were generally felt to be too time-consuming to become popular. However, with recent advances in natural language processing (NLP) and machine learning–assisted text analysis tools, this approach is worth revisiting. Automated approaches promise fast analysis of large bodies of text. The choice of analysis type depends on the answers one seeks to gather from the text, and there are more than can be covered here, but, for instance, frequency tables for commonly used terminology are useful, and sentiment analysis/opinion mining, or emotional analysis, can be automatically undertaken using AI tools (Collins and Johnston, 2023). See chapter 9 for more on using quantitative methods on qualitative data.

Projective techniques assume that when presented with ambiguous stimuli, we will project our own beliefs and attitudes onto those stimuli. Projective techniques involve presenting users with a stimulus, such as a product or sound, and asking them to respond in a way that reveals their underlying thoughts, feelings, or beliefs. In this way, we might understand opinion, motivation, or feeling beyond what users are usually able to articulate. They are indirect methods, meaning they do not ask respondents directly about the subject of interest. Instead, these techniques encourage respondents to project their subconscious thoughts onto a neutral stimulus. Techniques include, for instance, word association ("What comes to your mind when I play this sound?"), sentence completion, picture interpretation or drawing ("Draw this sound"), and word association tasks. For instance, a series of cards with attribute words or images can be created, and the participant is asked to choose which cards they associate with the sound. In this way, we can get results similar to free response techniques. The results are highly subjective, not statistically reliable, and time consuming (Vidal et al., 2013), but can be subjected to the AI techniques described above for free response methods.

6.3.1 Examples of Research Using Free Response Approaches

Altinsoy and Jekosch (2012) used free verbalization to generate a set of terminology (the semantic database) of vehicle sounds. In this case, they used an interview technique to elicit the terms. The study highlights the importance of understanding customer perceptions, interpretations, and expectations to evaluate sound quality effectively. The researchers developed a semantic space for vehicle sounds, which involved identifying numerous attributes related to the perception and quality of these sounds. This semantic space serves as a foundation for creating attribute scales, known as semantic differentials, which help in describing vehicle sound quality in a language that resonates with customer. The methodology used included a free verbalization interview, testing participants' understanding of acoustic attributes, and evaluating the ability of these attributes to describe perceptible vehicle sound properties. This approach led to the creation of a semantic database of twenty-eight attributes to describe vehicle sounds, ensuring the relevance and non-redundancy of the attributes.

Also working on car sounds, Nosulenko et al. (2013) used free verbalization to elicit a dataset of words associated with car sounds. The research involved evaluating and comparing the noises of seven diesel cars running at idle. Participants were asked to assess the differences in noise, identify the most appealing sound, and provide verbal explanations for their choices. They combined the method with paired comparisons of sounds. Interestingly, they found no difference between experts and naïve listeners in terms of response, but they did find a difference between men and women respondents.

6.4 Sorting, Ranking, and Projective Mapping

Sorting methods ask participants to group objects based on various criteria. There are a variety of sorting methods that can be adapted for sound-based research. Sorting tasks are very quick to administer and intuitive to participants, and large numbers of sounds can be handled in one session (Susini et al., 2012). Similarly, ranking tasks can be done quickly, and ask participants to just put sounds in a particular order or rank. There are many different types of sorting and ranking tasks. Some of the most common are as follows.

Ranking by elimination is a method to determine the ranking or order of items by successively eliminating the weakest or least preferred sounds until

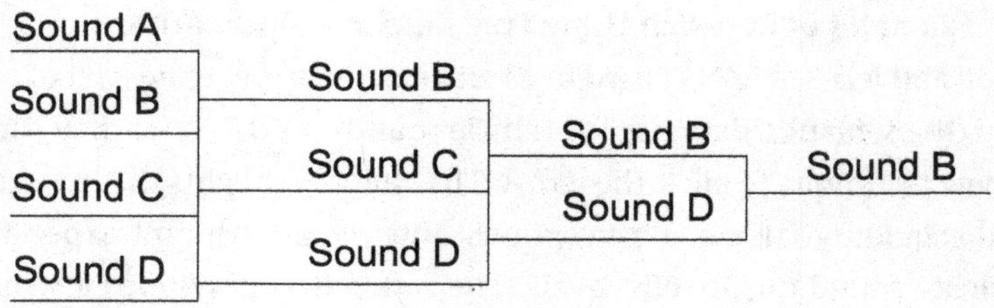

Figure 6.11
Example ranking-by-elimination test.

a ranking emerges. In each round, the least preferred sound is removed; this can be based on votes, scores, comparisons, or any other criteria. The last item to be removed in the process will be at the top of the ranking, and the first item removed will be at the bottom (figure 6.11).

Rank order can also be used to evaluate a set of sounds. It is a fast method, and is an effective way to quickly screen out unfavorable sounds. A set of sounds is presented to each participant, who arranges them according to some specified desired attribute. These attributes can be decided in advance or left to the participant (in a free sorting task). The maximum number of sounds to use to be effective is about ten, and usually closer to five. The participant is played all sounds to start, and then can re-trigger the sounds one at a time to decide on a ranking. They can play the sounds multiple times to refine their ranking. Data are typically analyzed using a Friedman test, and presented with average ranks for each sound.

Card sorting is less commonly used in studies of sound, since only digital cards are functional for sound. Participants are given a set of cards, each representing a sound, and are asked to group these cards into categories that make sense to them. This is useful for designing or evaluating information architecture and hierarchies. It's possible to use an open card sort task, in which participants can create their own categories to sort into, or a closed-card sort, which focuses on predefined categories.

Free sorting exercises are useful to help create attributes or lexicons for other tests. There is no rating of attributes, only the selection of attributes that each participant uses to describe the sounds. This method can help us determine what attributes are important to participants, who can create as many categories as they see fit. One difficulty with sonic material is, unlike visuals or objects, where all content can be viewed at once, sounds have to be played sequentially, and this may take the participant many tries to determine what categories should exist. The results are usually computed from similarity grouping using multidimensional scaling. For instance, the approach has been used on musical timbre (McAdams et al., 1995). A form of ranking procedure, called *random access*, has also been used for sound evaluation. The participant moves sound icons on a screen for each of the sounds into a rank with respect to sound quality (see, e.g., Fastl, 2005).

Best–worst scaling (BWS), including maximum difference scaling (MaxDiff), is a choice-based method that allows researchers to determine the relative importance or preference for multiple items by having participants identify the best and worst items within a given set (or the most and least important, most and least appealing, or other categories, selecting the parameters on which to judge each time). This way, we can rank a selection of sounds in terms of particular desirable attributes. If a product team has a list of potential features for a new product or an update to an existing product, BWS can help to prioritize which features are most and least preferred by users.

Participants are presented with a set of sounds. From each set, participants are asked to identify the sound they find most preferred (or most important or most helpful) and the sound they find least preferred (or least important or least helpful). This process is repeated with different combinations of sounds. By analyzing the aggregated responses, researchers can derive a ranking of sounds based on their relative importance or preference. In terms of sound, BWS has been used to ask participants to determine the best and worst sound based on a concept (Rosi et al., 2023).

With *projective mapping*, also known as *free choice profiling*, data are differentiated from other methods of sorting by a coordinate system, and therefore a position can be assigned to each product on a sheet of paper. Projective mapping is simple and intuitive for panelists as it doesn't require extensive training. It provides a broad overview of product landscape and perceived differences. It can be applied to various types of products and allows for open-ended evaluation without constraining panelists to specific attributes. It's a relatively fast method, especially when comparing a large number of sounds. However, while it provides a broad overview, it might not capture detailed attributes. One of the disadvantages is that all of the samples have to be presented at the same time, which is difficult for sound. It's possible, however, to allow participants to repeat sounds at will as they categorize them.

Projective mapping is usually used with trained panelists (ten to fifteen). Panelists are presented with a set of sounds, typically without any specific instructions about which attributes to focus on. Panelists position the products on a two-dimensional plane (the map) according to their perceptions. They do this based on any sensory attribute or combination of attributes that they deem important. Sounds placed closely together are perceived as similar, while those placed further apart are perceived as different. Each tester directly produces an individual map, or a similarity plot. The coordinates of each product's position are recorded for each panelist. Data evaluation is complex, and is typically carried out by means of hierarchical multiple factor analysis, multidimensional scaling, or generalized Procrustes analysis. There are also different versions of the projective mapping procedure, including napping and flash profiling. Flash profiling is a rapid method to gather and analyze perceptual descriptions and ratings of sound. For instance, let's say we have five sounds:

Sound A: Alarm beep

Sound B: Doorbell chime

Sound C: Dog barking

Sound D: Rainfall

Sound E: Phone ringtone.

We would play the sounds, and ask participants to freely describe them in their own words, noting down any attributes that come to mind. We collect all the descriptors from all participants. For example, descriptors might include: loud, soothing, annoying, melodious, sharp, or continuous. Then we give the participants a list of the compiled descriptors, and ask them to rate each sound on a scale, for instance, a seven-point Likert scale for each descriptor. To analyze, we calculate mean ratings for each sound on each descriptor (table 6.3). Then, we conduct a principal component analysis on the rating data to reduce dimensionality and identify the main factors driving perceptions of the sounds. Finally, we plot the sounds in the reduced space to visualize perceptual differences and similarities. Spider or radar plots for each sound can visualize their profiles based on the mean ratings for each descriptor, to give a quick look at the data to compare, for example, loud versus soothing for each of the five sounds (figure 6.12).

Table 6.4 shows an example of mean ratings for sounds.

Table 6.4
Example Mean Ratings for Sounds

Sound	Loud	Soothing	Annoying	Melodious	Sharp	Continuous
A	6.67	1	6.33	2.33	6.33	2.67
B	3.33	5.33	2.33	5.33	3.67	4
C	6.33	2	5.33	1	6.67	1.67
D	1.67	6.67	1	4.67	1.33	6.33
E	5.33	3.33	4.33	3	5	3.67

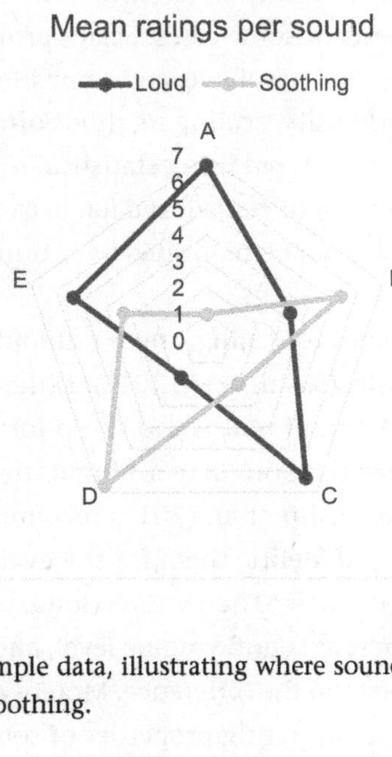

Figure 6.12
Radar graph of the example data, illustrating where sounds fall in terms of the two attributes of loud and soothing.

6.4.1 Example Uses of Ranking Methods to Evaluate Sound Design

Giacalone et al. (2017) used projective mapping to evaluate responses to twenty-five different types of loudspeakers, testing values on both expert and naïve assessors, and comparing methods as part of a broader investigation into the effectiveness of various sensory methods in evaluating sound quality. The study compared traditional methods like Quantitative Descriptive Analysis (QDA) with newer, rapid sensory methodologies such as projective mapping and check-all-that-apply (CATA). The study found that projective mapping and CATA, when conducted by experts, can yield results comparable to QDA, although they may be slightly less accurate. However, when projective mapping is performed by consumers, the results are less comparable to QDA. They found these methods to be more efficient than traditional sensory analysis, which requires more time and resources.

6.5 Rating Scales

Rating scales involve having users rate or rank their perceptions. There are a great many different types of scaling systems that can be used for user experience. Many scales are incorporated into other methods, but can also be used as standalone tests in some cases. Scales provide participants with some attribute or dimension to evaluate, and some form of scale showing an order of magnitude or intensity. Scaling methods often produce numerical data, which can be easily analyzed using statistical methods, and therefore can satisfy executives who want statistical data. Because they use standardized scales, these methods can be replicated over time to compare changes in product sound.

However, when it comes to scaling, the terminology of descriptors can significantly impact results (Susini et al., 2012). Different users might interpret scales differently: A sound that is too sharp for one user may not be sharp enough for another, but both may interpret the sound as somewhere in the middle of a scale. Susini et al. (2012) recommend using terminology in a sentence to help to define them for the evaluator. They describe, for instance, how the statement "The TV is too loud; we can't have a discussion" refers to the loudness as sound volume level, and not unpleasantness, which helps raters understand the difference. Moreover, they point out that users can feel uncertain about rating properties of sounds unless presented with a comparison or the entire range of sounds overall. Clear antonyms on differential scales (such as with a semantic differential) are important; finding a bipolar adjective pair can be difficult and confusing. One approach to overcome the problem may be to use a negation for the opposite term: for instance, sharp/not sharp (Kendall and Carterette, 1993). Susini et al. (2012) recommend that all sounds be presented for each semantic descriptor, rather than each sound being evaluated separately (i.e., all sounds rated for sharpness, so that users can compare). Whereas semantic scales compare stimuli along directly described dimensions, the number of scales required to assess a sound can be high, and semantic descriptors may not always be perceptually relevant (Susini et al., 2012).

The other issue with scales is that while they can tell us what users feel about sounds, they don't provide the reasons behind the ratings. Subtle differences, such as playing a sound too loudly, might result in a participant rating the sound lower than they would if the volume was appropriate to the context. As such, careful consideration and standardization of sounds is critical, such as normalizing all files to the same volume and setting the same volume for all participants. A standard reference sound file can also be used, so that participants are comparing the sounds to one file.

Multi-item measures consist of a list of emotion-related adjectives that the participants are required to check off or rate a sample against. Examples are the Multiple Affect Adjective Check List (MAACL) and Positive and Negative Affect Schedule (PANAS) (see Susini 2000 for an overview).

6.5.1 Semantic Differentials

The semantic differential scale is a type of *n*-point (typically five-, seven-, or nine-point) rating scale used to measure the connotative meaning of objects, events, and concepts using opposite pairs, such as smooth–rough, pleasant–unpleasant, soothing–jarring. The verbal descriptors are used as extreme values. The respondents are asked to choose where their response lies, between the two opposite adjectives. The closer to either extreme at each end, the higher the agreement (Hülsmeier et al., 2014).

The difficulty in using semantic differential scales is the selection of attributes or terms, since these vary for each sound stimulus. The scale is adapted to each specific sound under analysis, in other words, and can measure sensory properties (smooth versus rough), psychological or hedonic attributes, or emotional attributes. Terminology can vary by language and culture. As such, considerable time must go into developing the scale, and this process may involve other methods, such as free verbalization, to define what the salient attributes are. Pre-test practices should also be conducted.

For instance, we might create a seven-point scale with bipolar adjective pairs of:

- Pleasant–unpleasant
- Soft–loud

Table 6.5
Example Semantic Differential Rating Table from a Fictitious Experiment

Sound	Pleasant–Unpleasant	Soft–Loud	Relaxing–Stressful	Melodious–Jarring	Smooth–Harsh
A	6	7	7	6	6
B	2	4	3	3	3
C	5	6	6	5	5
D	1	2	1	2	1
E	4	5	4	4	4

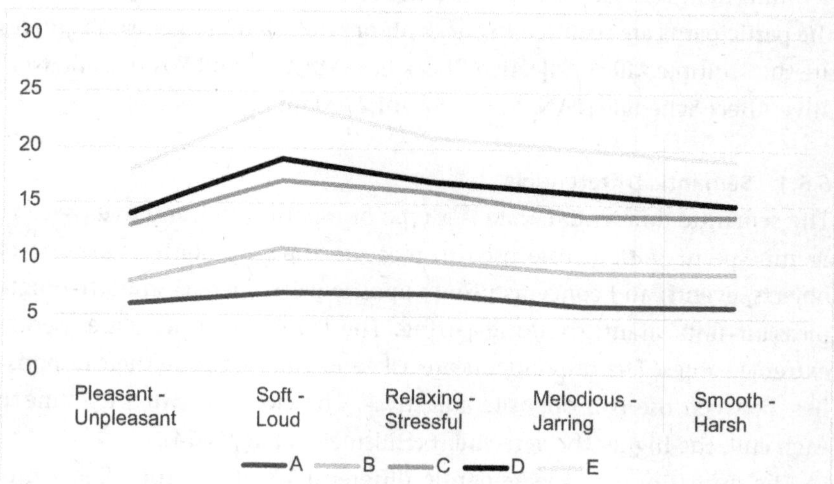

Figure 6.13
Example line chart for semantic differential presentation results.

- Relaxing–stressful
- Melodious–jarring
- Smooth–harsh.

Participants rate all the sounds on the scale, and we might end up with a table like that in table 6.5. From there, we can calculate the mean rating for each sound on each bipolar adjective pair. We can plot the means in a line chart, for example (figure 6.13).

The method of *verbal attribute magnitude estimation* (VAME) is a variant of the semantic differential method that might improve the interpretability of the results. With the VAME method, bipolar scales are not defined by opposing semantic descriptors but by one semantic descriptor and its negation (e.g., not loud and loud) (see Lipscomb and Kerins, 2004).

6.5.2 Repertory Grid Method

The Repertory Grid Method, also known as Repertory Grid Technique, is a method developed in the 1950s and uses ideas of *personal constructs*, the mental categories or dichotomies that people use to distinguish and make sense of the world around them. For example, in the context of product evaluation, someone might use constructs like cheap versus expensive or healthy versus unhealthy to assess different products.

Repertory Grid consists of several stages: First, the terminology is developed—and this can be done using a variety of methods discussed in this chapter (such as free response or sorting) that can help to define a list of attributes to test for. Alternatively, by allowing participants to define their own attributes, we can also see what aspects of the sounds are most important to them. The participant then describes the two polar opposites for the end points on the grid, which consists of three, five-, or seven-point scales.

In the second phase, participants rate each of the stimuli based on the constructs they created in the first part. The data are then used to create a grid, where attributes are listed along the top row and down the first column, and participants' constructs are used to fill in the cells. Each cell in the grid represents the participant's assessment of how the two elements are related according to a specific construct. This grid becomes a visual representation of the participant's cognitive structure. Researchers can analyze the grids in various ways, such as through factor analysis or cluster analysis, to identify patterns and relationships among the constructs and elements. This analysis can reveal common themes, shared constructs, and differences among participants.

Generally, between eight and fifteen participants are used, with between five and ten stimuli (Grill et al., 2011). There is little to no training required, and naïve participants can be used. However, data analysis is not straightforward, usually using hierarchical cluster analysis and non-hierarchical cluster analysis.

For example, let's say I have five sounds:

Sound A: Alarm beep

Sound B: Doorbell chime

Sound C: Dog barking

Sound D: Rainfall

Sound E: Phone ringtone.

To help to elicit constructs to evaluate, I may use what is called triadic elicitation: I present participants with sets of three sounds (triads) and ask them to identify a way in which two of the sounds are similar and different from the third. This helps to elicit personal constructs.

For example, presenting the triad alarm beep, doorbell chime, and dog barking, we may get: "The alarm beep and doorbell chime are more mechanical, while dog barking is natural." Then, collect all of the constructs and create bipolar scales. Gather constructs from several participants to compile a comprehensive list of constructs.

Example constructs:

- Mechanical versus natural
- High-pitched versus low-pitched
- Continuous versus intermittent
- Pleasant versus unpleasant
- Attention-grabbing versus subtle.

We then have participants rate all of the sounds on the scales for the constructs. Then we calculate the mean scores across participants. Results are often presented as a heatmap or grid plot (table 6.6).

Table 6.6
Example Heatmap Presentation of Results from Fictitious Experiment

Construct	Alarm Beep	Doorbell Chime	Dog Barking	Rainfall	Phone Ringtone
Mechanical–Natural	2	3	7	6	3
High-Pitched–Low-Pitched	1	4	5	7	3
Continuous–Intermittent	6	2	4	1	3
Pleasant–Unpleasant	6	3	5	2	4
Attention-Grabbing–Subtle	1	4	5	6	3

6.5.3 Absolute Category Rating, Degradation Category Rating, and Comparison Category Rating

In *Absolute Category Rating* (ACR), participants are presented with a single stimulus (e.g., a sound file) and are asked to rate its quality based on a predefined scale without comparing it with any reference or other stimuli. Typically, there is a standardized five-point scale on which users rate the sample, for instance, from bad to excellent. However, the number of points and the labels can vary based on the specific test. It is often used in situations where a quick and direct assessment of perceived quality is desired. The advantage of ACR is its simplicity. Since participants are presented with only a single sound and do not have to make comparative judgments, the test procedure is straightforward. However, one of the method's limitations is that it doesn't provide detailed information on specific quality differences between stimuli. In multimedia quality assessment, ACR has outperformed other evaluation methods (Strohmeier et al., 2010).

Example: Rate the sound of the button press on this coffee machine. Use the following scale:

1: Bad

2: Poor

3: Fair

4: Good

5: Excellent

Summary example results: Mean rating: 3.7; mode: 4; standard deviation: ~0.9. Interpretation: The button press sound is generally perceived as good, though a few participants rated it lower, suggesting room for improvement in perceived quality or pleasantness.

Unlike Absolute Category Rating where stimuli are rated in isolation, *Degradation Category Rating* (DCR) involves comparing a degraded version of a stimulus with its reference (usually a high-quality, undegraded version) and then rating the perceived degradation. This method is useful when the aim is to gauge the perceived degradation caused by specific factors, such as data compression or technological choices (e.g., speaker options in a product). One of the main advantages of DCR over methods like ACR is that it can provide a more direct measure of the impact of specific impairments on perceived quality. For example, do users notice the difference between the ten-cent speaker and the high-end speaker in the device?

Example: You will hear two sounds:

1. A reference sound (ideal version)
2. A test sound (possibly degraded).

Rate the difference in quality between the two sounds using the scale below:

5: Imperceptible

4: Perceptible but not annoying

3: Slightly annoying

2: Annoying

1: Very annoying

Summary example results: Mean rating: 3.7; mode: 4; standard deviation: ~0.9. Interpretation: The degradation between the reference and test sound is generally perceptible but not annoying, indicating a small but acceptable change in sound quality.

Comparison Category Rating (CCR) involves directly comparing two or more stimuli without a specific reference. In a typical CCR test participants are presented with two or more versions of a sound, and are then asked to compare the sounds and rate one in relation to the other(s) based on a predefined scale. The scale might be designed to assess, for instance, preference (e.g., "Prefer sound A much more than sound B") or quality difference (e.g., "Sound A is much better in quality than sound B"). The nature of the scale depends on the objectives of the test. CCR is useful when there isn't a clear high-quality reference or original sample.

Example test question: You will hear two sounds, A and B, played in random order. Rate the quality of B compared with A using the following scale:

+3: Much better

+2: Better

+1: Slightly better

0: About the same

−1: Slightly worse

−2: Worse

−3: Much worse

Summary example results: Mean rating: +0.6; mode: +1; standard deviation: ~1.2. Interpretation: On average, participants perceived the new button sound (B) as slightly better than the original (A). However, the responses show variability, suggesting individual differences in preference or sensitivity.

6.5.4 Magnitude Estimation and Category-Ratio Scales

Instead of relying on fixed points or categories, as with many traditional scales, magnitude estimation requires participants to assign numbers to stimuli based on the *perceived magnitude* of an attribute, relative to a reference or standard. Participants are first presented with a reference sound and are given a numeric value to associate with its perceived magnitude. Participants are then presented with other sounds and asked to assign numbers to these based on their perceived magnitudes in comparison to the reference sound. For instance, imagine an experiment where participants are asked to evaluate the loudness of sounds. They could be presented with a sound at a moderate volume and told it has a value of 10. Then, when exposed to a sound they perceive as twice as loud, they might assign a value of 20.

Unlike fixed-point scales, participants are not restricted to a predetermined set of response options. The method can reveal subtle perceptual differences and extremes that might be glossed over with other scales (participants tend to avoid extremes when estimating, termed *regression bias*). However, the method can be more complex for participants, especially if they are unfamiliar with the concept. Some participants might not use a consistent scale throughout their evaluations. For instance, the ratio of perceived differences might change as they proceed through a large group of sounds. Data from magnitude estimation can also be more challenging to analyze because the data don't adhere to fixed intervals or categories. Since different participants may use different scales (e.g., one person using numbers in the tens while another uses numbers in the hundreds), it's often necessary to normalize or standardize responses. One common approach is to divide each response by the participant's value for the reference stimulus. This creates a ratio or proportion relative to the reference, making data more comparable across participants.

Like magnitude estimation scales, *category-ratio scales* allow participants to assign their own numerical values to stimuli. Participants are given a scale with predefined categories, each associated with a numerical ratio that reflects an intensity level. For example, a sound might be rated on a scale from 0 to 10 where each number is linked to descriptors such as no sound at 0, very weak at 1, weak at 3, moderate at 5, strong at 7, and very strong at 10. The participants select a category that matches their perception of the stimulus intensity. This method tends to produce ordinal data, but with the assumption of equal intervals implied by the numerical ratios, it can approximate interval-level information. In the context of user experience, category-ratio scales can be used in situations where it's crucial to capture the intensity or magnitude of perceptions, feelings, or experiences, and where the ratio properties of the scale can provide more in-depth insights.

Magnitude Estimation Test Example Imagine we are testing how loud different coffee machine beeps sound.

Instructions: "You'll first hear a reference beep. This is assigned a value of 100. For each following sound, assign a number representing how loud it feels to you, relative to the reference. If a sound feels twice as loud, assign it 200. If it feels half as loud, assign it 50. You may use any positive number."

Play Standard Beep (100)

Play Sound A

Play Sound B

Play Sound C

Results: Sound mean magnitude estimation score: A: 80; B: 102; C: 172.

Interpretation: Sound A is perceived as softer than the standard (100). Sound B is perceived as about the same as the standard. Sound C is perceived as significantly louder, close to 1.7 times the standard.

However, the nonlinear nature of some category-ratio scales like magnitude estimation can be confusing to participants. With scales that have nonlinear progressions, participants might find it counterintuitive to have uneven jumps between numbers, which can impact the reliability of their responses. Analyzing data from category-ratio scales requires an understanding of the scale's inherent properties, notably, its ratio characteristics.

The original category-ratio scale was adapted to hedonic measurement in the labeled affective magnitude (LAM) scale. The LAM scale typically ranges from 0 to 100, with specific labels or anchors that denote various levels of affective intensity. These anchors help participants understand and quantify their emotional responses more accurately. In the context of UX sound research, the LAM scale can be used to evaluate how different sounds affect users emotionally. For example, participants might be asked to listen to various sounds associated with a product and rate their affective response using the LAM scale. This can help researchers understand which sounds evoke the most positive or negative emotions.

Example of LAM scale use: Imagine a study where participants are evaluating the sound of different car engines. They might use the LAM scale to rate their emotional response to each engine sound. A sample question could be, "Rate your affective response to the sound of this car engine on the following scale." Participants would mark their response along this continuum, providing valuable data on how different sounds are perceived emotionally.

Example of LAM-Style Rating System

0	20	40	60	80	100
No affect	Slightly affective	Moderately affective	Very affective	Extremely affective	Maximum affective response

6.5.5 Comparative Scales and Comparative Intensity Scales

Comparative intensity scales involve a scaling technique where participants are asked to assess the relative intensity or magnitude of various sounds or attributes by comparing them against each other. Unlike direct scaling methods that evaluate items separately, this approach requires consideration of sounds in comparison to others. For example, participants might be asked to rate the intensity of each sound presented in terms of loudness or softness in comparison to a reference. Typically, items are presented in pairs, and respondents are asked to identify which sound in the pair exhibits greater intensity or preference for a specific attribute. Participants may also be asked to rank items based on the intensity of a particular attribute, providing an ordinal scale of preference or intensity without quantifying the differences between ranks. Advantages of this method include the ease of making relative judgments, which are often more intuitive than absolute assessments, and the potential reduction of some biases through direct comparisons. However, the approach has disadvantages as well: It does not yield absolute measurements but only relative differences, and because judgments are relative, it is hard to determine if perceptions would shift with the addition or removal of certain items from the comparison set.

Example Comparative Intensity Scale Scenario: You're evaluating how urgent different coffee machine alert sounds feel.

Instructions: "For each pair of sounds, indicate which one feels more urgent, and by how much. Listen to sound A and sound B (etc.), then choose. Which sound feels more urgent?"

( ) A: Much more urgent

( ) A: More urgent

( ) A: Slightly more urgent

( ) About the same

( ) B: Slightly more urgent

( ) B: More urgent

( ) B: Much more urgent

(repeat for A and C, and then B and C).

Interpretation: Sound C is consistently judged as more urgent than both A and B. Sound B is perceived as slightly more urgent than A, but less than C. Sound A is generally the least urgent.

6.5.6 Hedonic Scale Method

Hedonic scales are used primarily in sensory evaluation to measure the degree of liking or disliking of a product. These are sometimes referred to as affective tests or affective scaling. A typical hedonic scale might range from "like extremely" to "dislike extremely," with various gradations of liking or disliking in between. The nine-point hedonic scale is the most common for testing user preferences, although five- and seven-point scales are also used (e.g., figure 6.14). It is a balanced bipolar scale, with neutral in the middle, and four positive or negative categories.

Table 6.7
Example Results from Comparative Intensity Scale Test (N=8 participants)

Pair Compared	Much More Urgent	More Urgent	Slightly More Urgent	About the Same
A vs. B	1 (B)	3 (B)	2 (B)	2
A vs. C	2 (C)	2 (C)	3 (C)	1
B vs. C	0	2 (C)	3 (C)	3

Figure 6.14
A five-point hedonic scale with emojis.

Hedonic scaling offers a straightforward way for respondents to express their level of like or dislike. However, perceptions of pleasure or preferences can vary across cultures, which can influence the way products are evaluated on a hedonic scale. Like other unidimensional scales, hedonic scales focus on a single aspect (liking/disliking). They don't provide insights into *why* a sound is liked or disliked or other nuances of the experience. Moreover, participants often avoid using the extremes, narrowing results around the middle. A large sample size (over seventy-five responses per stimulus) is necessary to approximate normality in order to make valid statistical inferences (Lim, 2011).

6.5.7 Just-About-Right Scale

The just-about-right (JAR) scale is used primarily in the food and beverage industry, but it can also be applied to sound. The scale is employed to gauge consumer perceptions about specific attributes to determine whether these attributes are at the desired or optimal level (the "goldilocks zone"). The goal is to determine whether an attribute needs to be adjusted to meet consumer preferences. If a significant portion of respondents indicates that the volume is "too much," then the product development team might consider reducing its loudness.

For instance, a sound might be evaluated based on attributes like sharpness or loudness. Respondents are asked to rate these attributes on the scale, which typically includes options such as too little, just about right, and too much. The scale generally uses these three points, but can be expanded for additional nuance.

Despite being less sensitive to subtle differences, making it less suitable for detailed analysis, it's valuable for initial screenings and setting product benchmarks. Analysis typically involves calculating the percentage of responses in each category, with adjustments considered necessary if a significant deviation from the just-about-right category is observed. This approach ensures that products align closely with consumer preferences, guiding necessary changes based on the distribution of responses.

6.5.8 Examples of Rating Scales Applied to Sound Design

Dickson and Bolin (2014) explored the use of a continuous judgment by category-ratio (CJCR) scaling method to assess the annoyance caused by aircraft noise. Participants rated their annoyance on a Borg CR 100 scale during the playback of ten different flyover sequence sounds. The Borg CR scale is a ten-point perceptual strength scale where participants rank responses from very weak to extremely strong. This method allowed for the creation of a temporal stream of annoyance levels throughout the entire flyover sequence, which could be further analyzed for patterns and insights. The use of a category-ratio scale, as opposed to a simple category scale, enabled mathematical operations such as averaging and provided higher resolution in the responses. The study found notable differences in perception during the time segments where the sound was modified. Additionally, the temporal annoyance data were converted into overall judgments of the sounds, which were consistent with previous findings obtained through semantic differential and paired comparison methods.

Altinsoy and Jekosch (2012) used a semantic differential test to explore customer perception of vehicle sounds, emphasizing that traditional physical and psychoacoustical measures are insufficient for capturing the nuances of customer perception. Instead, the study highlights the importance of understanding customer perceptions, interpretations, and expectations to evaluate sound quality effectively. They first used a free verbalization interview, then used the words to create the semantic differential scale, developing two sets of twenty-eight attributes tailored for interior and exterior driving conditions, providing a structured framework for assessing vehicle sound quality from a customer-centric perspective.

Grill et al. (2011) used the repertory grid method to explore textural sounds (what they defined as sounds that appear stationary, as opposed to evolving over time). The study aims to enhance the design of intuitive interfaces for navigating extensive sound libraries. By using a mixed-methods approach within the repertory grid framework, the researchers identified ten bipolar perceptual qualities relevant to textural sounds. A subsequent web-based study was conducted to measure inter-rater agreement and the mutual similarity of these perceptual qualities using a selection of 100 textural sounds. Additionally, initial experiments were performed to examine the correlation between standard audio descriptors and the identified perceptual qualities. The paper details the process of coding the concepts evoked.

6.6 Temporal Techniques

Temporal methods focus on understanding how user experiences or perceptions change over time. They help researchers capture the dynamic nature of user experiences, and identify key moments or factors that influence these changes.

The *experience sampling method* sends prompts to users at random or predetermined intervals during a defined time period to ask questions about their current situation as related to a product. Participants report on their thoughts, feelings, behaviors, and environmental context at various times throughout the day. This method aims to capture experiences in real time or near real time, thereby reducing recall bias and providing a more accurate representation of daily life. Because feedback is collected multiple times per day over a length of time, researchers can gain a deeper understanding of the context and fluctuations in participants' experiences over time.

However, being prompted multiple times a day can be disruptive or burdensome for participants, potentially leading to dropout or inconsistent responses. Collecting data multiple times per day from numerous participants can also result in a large volume of data, which can be challenging to manage and analyze.

Another test is the *descriptive experience sampling* (DES) method, which is very similar, but its aims are to capture a narrower range of thoughts, feelings, and sensations, whereas the standard experience sampling method tends to include mood, stress levels, and more contextual information (Hurlburt and Akhter, 2006).

The *temporal dominance of sensations* (TDS) test also comes from food studies and is designed to study the dynamic profile of sensations perceived over time during the consumption of a product. A product is presented to a panelist (usually trained or at least familiar with the product category). As the panelist consumes the product, they use an interface (often computer-based) to indicate which sensation or attribute is dominant at any given moment. With the food tests, this could be, for example, sweetness, bitterness, crunchiness, or creaminess, depending on the product. The test continues for a specified duration or until the product is fully consumed. Over time, and across multiple panelists, the TDS test produces a temporal profile, showing which sensations dominate and at what stages of consumption. This is typically undertaken with trained panelists and requires specialized software. Attributes can be checked multiple times or ignored if they are not present. The time of the click on the attribute is recorded. Although not directly used to measure *only* sound, the method has been used to measure sound's impact on taste, and could be adapted to measure responses to sound over time. An adaptation of the TDS method, *temporal dominance of emotions* focuses on tracking dominant emotions over time. Participants adjust a dial or slider continuously to indicate their feelings or emotions during an experience, such as watching an advertisement. This method is particularly useful for products where emotional responses evolve over time.

6.6.1 Example Use of Temporal Techniques

Juslin et al. (2008) used an experience sampling method to explore emotional responses to music as they occurred in everyday life. Participants, who were college students, carried a mobile device that prompted them seven times per day at random intervals over a two-week period. Each time they were signaled, they completed a questionnaire on the device. The study found that music was present in 37 percent of the episodes recorded, and in 64 percent of these music episodes, participants reported that the music influenced their emotions. The emotions most frequently associated with music included happiness/elation and nostalgia/longing, while nonmusical episodes were more often linked to emotions such as anger/irritation and anxiety/fear. The study highlighted the importance of considering the listener's personality and the situational context in understanding the prevalence and nature of musical emotions. The researchers were able to show the influence of music on affective state, as well as track location and what the participant was doing at the time.

Galmarini et al. (2021) used temporal dominance of sensations and emotions to test how music altered the perception of coffee over time using the temporal dominance of sensations and temporal dominance of emotions methods. The research involved two groups of consumers evaluating two different commercial filter coffees under silent conditions and while listening to music with "sweet" and "bitter" connotations. The findings revealed that musical stimuli significantly altered coffee perception. Specifically, the duration of the bitter sensation increased with bitter music and decreased with sweet music. Conversely, the sweet attribute, rarely chosen in silence, became more dominant with sweet music. Music also had a pronounced effect on emotions, with sweet music evoking joy, surprise, and some sadness, while bitter music was associated with anger and fear. This study highlights the concept of *sonic seasoning*, where music influences the sensory and emotional experience of a familiar product like coffee.

6.7 Threshold Techniques

The *go/no-go method* is a type of threshold testing used to determine the detection threshold of a particular sensory stimulus, such as sound. It's a simple, forced-choice procedure where participants are presented with a stimulus and must decide whether they can detect the stimulus (go) or not (no-go). This method is often used when the researcher is interested in the presence or absence of a sensory attribute, rather than its intensity or degree. For each sound, the participant must respond whether they detect the attribute or not. There are no options for uncertain responses, forcing a binary decision. The detection threshold is determined as the lowest concentration level at which a predetermined percentage of participants (often 50 percent) can correctly identify the presence of the stimulus more often than chance (Godfrey et al., 2023). The go/no-go method is beneficial because it is straightforward and quick, making it suitable for testing with a large number of participants. It also minimizes bias because the participant is not required to grade the intensity of the stimulus, only to report its presence or absence.

6.7.1 Example of Threshold Techniques Used for Sound

Zhelezoglo et al. (2021) used go/no-go matching to sample procedures to compare the perception of visual objects with an auditory stimulus. In this procedure, participants were presented with a sample and a comparison stimulus sequentially. The training involved presenting an auditory stimulus followed by a visual comparison, where participants earned points by touching the correct visual comparison related to the auditory sample. Incorrect responses or refraining from touching unrelated comparisons resulted in no consequences. After training on AB/AC relations, participants were tested on untrained relations (BA/CA and BC/CB), as well as on their ability to label and sort stimuli. The results from two experiments showed that all participants met the emergence criteria for untrained relations and sorting tasks.

Kwa et al. (2024) explored the development and impact of an audio-guided deep breathing exercise combined with an auditory go/no-go task on the wellness of young adults. The research aimed to assess the immediate effects of this intervention on a healthy population by measuring changes in tidal volume, attention levels, and psychological well-being. The study involved thirty-two healthy young adults who participated in a five-minute audio-guided deep breathing session using natural sounds. The participants' attention levels were evaluated using an audio-based go/no-go task developed with Arduino technology. Psychological assessments were conducted using the Rosenberg Self-Esteem Scale (RSES), Cognitive and Affective Mindfulness Scale—Revised (CAMS-R), and Perceived Stress Scale (PSS). The results indicated a significant increase in tidal volume and task reaction time post-intervention ($p = 0.003$ and $p = 0.033$, respectively). Additionally, significant correlations were found between task accuracy and commission errors, as well as between CAMS-R scores and various task performance metrics. The study concludes that the audio-guided deep breathing and auditory go/no-go task may enhance young adults' well-being and suggest the potential of using the go/no-go task to assess attention in different groups while promoting physical and mental health.

6.8 Descriptive Analysis

Descriptive analysis testing methods have developed from sensory evaluation and focus on gathering detailed descriptive information about a product or experience, through helping to identify, analyze, evaluate, and describe the various sensory attributes of a product or service (see Kemp et al., 2018). Descriptive analysis aims to answer the questions "What is the nature of the difference between these two products?" or "How do users describe this product?" or "Which sound attributes (e.g., pitch, volume, timbre) are most important to our customers, and how can products be adjusted to align with those preferences?" or "Is the product's sound consistent across multiple uses or different units of the same product?" or "Are there noticeable variations in sound between different models or versions of the product?"

These methods can be useful to develop a set of terminology (a lexicon) that can then be used in other types of testing on a wider audience, such as to create attributes for a semantic differential scale. The data can also be used to fingerprint a product for comparison with later versions or other versions of the product.

Descriptive analysis approaches have grown from food science, where the flavor profiling method has been used for decades. New methods have developed more recently, with some general standards for descriptive analysis having been developed (e.g., Gacula, 1997). These methods have been used for very limited, specific aspects of sound, such as low-frequency room acoustics (Wankling et al., 2012), and spatial audio displays (Zacharov and Koivuniemi, 2001), although they could be applied more widely.

Descriptive analysis typically uses a small group (six to twelve) of trained assessors, and usually relies on an agreed-upon, standardized series of definitions for attributes to evaluate. Typically, these attributes are formed in a preliminary group session where all of the panel has input until a consensus is agreed. Often, reference standards are found to illustrate the property or attribute. Finally, an anchor term is chosen for the high and low ends of the scales to be used. Practice sessions are held, and then finally the panel evaluates each item according to the scale created in terms of the attributes present and their intensity. The terms are a considerable source of variability in test results if they are not consistently agreed upon, and the establishing of attributes to test is a critical stage of the process.

There are many different variations of descriptive testing that involve slightly different approaches, including classic testing, consensus profiling, qualitative descriptive analysis, free choice profiling, flash profiling, and audio descriptive analysis and mapping.

6.8.1 Classic/Conventional Descriptive Tests and Consensus Profiling

In a classic descriptive test, trained panelists describe the attributes that are to be focused on, and are trained extensively to identify and rate the intensity of each attribute. Adjectives used to assess and describe the product can be free-form or selected from a list, although no standard list exists for sound. The method is time-consuming and panelists must be carefully screened. Consensus profiling is very similar to classic descriptive testing, deriving from the original flavor profiling technique, but involves the collaboration of a group (six to twenty) of trained panelists to collectively describe the attributes of the samples being evaluated, and arrive at a consensus. The process is iterative, and typically between five and ten sessions are required to develop a consensus vocabulary. Since the test relies on consensus, statistics are not necessary to evaluate the results. The downsides of the consensus method are that it may lose individual perceptions and takes considerable training to use effectively.

6.8.2 Quantitative Descriptive Analysis

Quantitative descriptive analysis was developed in the 1970s for food evaluation, and uses a trained panel (ten to twelve) to describe all of the sensory properties of a product as with classic tests described above, with the addition of then rating the intensities of the attributes for the product. Working with a panel leader, assessors develop or modify a list of attributes to describe all of the products' properties in either classic or consensus methods described above. After agreeing on the list, the panel undergoes practice sessions of rating different samples to learn the parameters of what will be rated. Then, they complete a repeated-measures design (typically, a four-replicate design) rating, to compare their own results over time for consistency. The attribute evaluation is not intended to be absolute; instead, intensity measurements are only relative to the products being evaluated, so are pushed to their highest or lowest intensities.

6.8.3 Free-Choice Profiling and Flash Profiling

Free-choice profiling, also known as the evaluation grid approach, is a variant of qualitative descriptive analysis, using a trained or untrained panel, and focuses on liking or acceptance responses to each attribute, rather than an empirical rating of the attributes. Panelists (ten to thirty) develop their own individual terms to describe the product, and then rate them according to their own perceptions. Free-choice profiling therefore is useful for comparing two products, or when a general understanding is required in one-off studies, but is not suitable for measuring changes over time or when reproducible data are required. An advantage of the method is that no training is required, and a convenience sample will suffice. However, results are not reproducible, and the lack of a formal or defined lexicon makes comparisons between assessors difficult. Data are typically analyzed using the generalized Procrustes analysis method, which can be difficult to use.

Flash profiling was derived from free-choice profiling. With flash profiling, assessors also create their own terminology for attributes, and we can use naïve, semi-trained, or trained panels. However, with flash profiling, the samples are presented all at the same time, rather than one at a time. All samples are compared, to generate the best terms to discriminate between the samples. Then, product samples are ranked from low to high according to each descriptor. Instead of intensity on a rating scale, then, the products are compared with each other based on the attributes used. This makes it a fast method (hence the name *flash*), useful in early product development stages. However, although the test is quick, analysis is not, since everyone is using different terms. A generalized Procrustes analysis is typically used to develop a sensory map.

6.8.4 Audio Descriptive Analysis and Mapping

Zacharov and Koivuniemi (2001) developed a method for the quality evaluation of audio equipment and spatial audio displays called Audio Descriptive Analysis and Mapping (ADAM). The method is designed to establish the perceptual structure of a specific domain (such as in the original case, spatial sound reproduction). A language development task is performed with trained participants, and a consensus vocabulary is achieved that is used for attribute ratings by trained participants.

6.8.5 Examples of Descriptive Analysis Techniques Used for Sound Design

Benghanem et al. (2021) attempted a variety of methods to evaluate the best methods for vehicle sound evaluation. They first employed free verbalization to elicit terminology to create appropriate attributes to use for evaluation. A panel and online anonymous surveys refined the terminology to evaluate. A key finding was that loudness equalization of sound samples during listening tests results in more refined sensory profiles compared with non-equalized samples. This approach helps establish sensory profiles, desire-to-buy values, and desirable sound profiles for the vehicles tested, providing valuable insights for acoustical engineers in the design process.

Walton et al. (2017) combined preference ratings with sensory profiling to evaluate audio technology. They argue that by mapping user preferences to individually elicited attributes, a deeper understanding can be formed. They provide an overview of the Open Profiling of Quality (OPQ) method for sound, which was initially used in visual and audiovisual evaluations, and adapt it for audio reproduction systems. This method involves psychoperceptual evaluation, sensory profiling, and external preference mapping, making it suitable for naïve listeners. The paper presents a methodological case study detailing the implementation and adaptation of OPQ for comparing different audio reproduction systems, highlighting its potential to provide insights into how individual sensory attributes contribute to the perceived quality of audio experiences.

6.9 Conclusions and Take-Aways

There are many useful methods for exploring sonic user experience research. Existing methods that you may be familiar with when it comes to other UX testing can be adapted for sound with a little effort to answer your research questions. The wide array of methods available for studying sonic UX highlights the multifaceted nature of sound in user experience. Each method offers unique insights that contribute to a deeper understanding of how sound influences users. Since this is a very new area of research, however, there are no tried and true methods for any particular aspect of product sound design. This makes it all the more important to use more than one method where time and resources permit, since the results may require some extra steps in interpretation without established benchmarks. By using a diverse set of methodologies, researchers can capture a comprehensive picture of user interactions with sound, ensuring that designs are effective.

The choice of method should always be driven by the specific research question and context, considering factors such as the product type, target audience, and available resources. While challenges such as resource constraints and variability in user responses exist, these can be mitigated through thoughtful methodological design and a commitment to iterative testing.

There are methods that I have not covered here that can also be adapted for sound; for instance, user journeys or personas might be adapted to focus on sonic elements. An auditory user journey, for instance, might highlight user interactions with sound throughout their experience, and could be used to compare an experience with and without sound, and illustrate where personalization options may be warranted. I would encourage all readers to share their methods, the successes and failures, and the results of testing with the research community so that we may develop the methods further.

7 Psychophysiological and Neuroscientific Methods

One disadvantage of the subjective, self-report measures commonly used in UX research and detailed in the previous chapter is that people's introspective abilities are often quite poor, making the results questionable. Moreover, there is a response bias in that people tend to provide answers that they think are the best answers, rather than expressing their true feelings (van der Zwaag et al., 2011). Evidence suggests that conscious and unconscious *arousal* are independent (arousal is the excitement, stimulation, and bodily activation) and influence attitudes toward products in different ways (Bettiga et al., 2017). As well, research has shown that unconscious reactions to stimuli are more accurate indicators of actual thought and behavior than self-reports (Alexomanolaki et al., 2010). While most UX research treats arousal as a conscious response, psychophysiological research can show underlying, unconscious reactions of users (Bell et al., 2018).

Psychophysiological and neuroscientific measures can be useful to supplement (or in some cases even replace) self-report measures in UX research. These measurements provide valuable insights into our users' emotional, cognitive, and physiological responses during their interactions with sound. Psychophysiological measurements can be taken in real time during playback of a stimulus, and can capture emotions that would have otherwise been unnoticed. Psychophysiological responses are most often used in emotion-based research, but as will be shown, can also demonstrate the cognitive processes involved.

In this chapter, I break down the various ways of testing emotional and cognitive responses using psychophysiological and neuroscientific measures. The distinction between these relates to the concept that neuroscientific methods are more specific to measuring the brain and central nervous system, while psychophysiological methods encompass a broader range of bodily responses, including peripheral measures. Both approaches can be used complementarily in experiment design and/or with self-report methods. Each method has its own advantages and disadvantages. There isn't space here to go into too much detail, nor cover every possible method; rather, I introduce these as tools that may be a useful addition to your UX sound research, and provide information on how these approaches have been used in sound and music research.

These measures are useful to accurately answer questions about perception, such as whether a user notices or pays attention to a sound, whether they respond to a sound, how they respond to sound, and what emotions they may be experiencing. However, the methods are sometimes costly, complex, and require training to run and analyze. Moreover, the laboratory settings sometimes necessary for the evaluation can give rise to stress or arousal responses in many people, so experience with the equipment and analysis is necessary to screen out the resulting data noise introduced by these factors.

7.1 Measuring the Skin: Electrodermal Activity and Skin Temperature

Electrodermal activity refers to the electrical activity in the skin, and these activity signals can be measured. Electrodermal activity reflects changes in emotional arousal, making it a valuable tool for studying emotional responses. It can capture real-time fluctuations in arousal during events, stimuli, or tasks.

Electrodermal activity is closely linked to emotional arousal and stress. When individuals experience emotional responses such as excitement, anxiety, or stress, the activity of their sweat glands increases, leading to a decrease in skin conductance (and an increase in resistance). Electrodermal activity is therefore used to study emotional responses, since it measures sympathetic nervous system arousal without interference from parasympathetic activity. It can be used to help assess the emotional impact of products, interfaces, or sound on users.

Electrodermal activity is an umbrella term used to incorporate galvanic skin response (GSR, which specifically measures change in skin conductance that occur due to sweat gland activity), sympathetic skin response (which measures any kind of sympathetic arousal, not limited to psychological/emotional factors), and/or skin conductance levels (SCL, which can fluctuate over time and are usually measured over the long term). When it comes to UX research, GSR is the most common measure, used to study emotional responses, arousal levels, and stress reactions due to emotional or psychological stimuli (Zaki and Islam, 2021).

Electrodermal activity has been shown to be a useful measure of response to advertising stimuli (see Caruelle et al., 2019). More recently, electrodermal activity has been used to explore cognitive load processing, since more demanding tasks evoke increased electrodermal activity responses (Buchwald et al., 2019). Electrodermal activity has also been used to compare subjective and objective measures of emotional response to music (e.g., Williams et al., 2015). Juslin and Västfjäll (2008) have explored how music tempo (fast vs. slow) and mode (major vs. minor) influence physiological responses in listeners using skin conductance levels. They found that slow tempo and minor mode music elicited higher skin conductance, indicating increased arousal, compared with fast tempo and major mode music. These physiological responses corresponded with listeners' self-reported emotional ratings, suggesting a link between the skin conductance and emotional experiences induced by music. Some studies have shown that skin conductance levels are higher with fear and happiness, but also increase during unexpected harmonies (van der Zwaag et al., 2011).

As discussed in chapter 3, in my former lab at the University of Waterloo, we studied sound and music in slot machines. We had discovered that slot machines incorporated what we termed *losses disguised as wins*, where, for example, you put in a dollar and win back fifty cents, but get all of the visual and sonic cues of a regular win, even though you lost fifty cents. We used electrodermal activity to measure user responses to wins, losses, and losses disguised as wins. Skin conductance responses were similar for wins as they were for losses disguised as wins, indicating that the physiological response of players was basically being treated as (and miscategorized as) a win by players (Dixon et al., 2013).

The equipment to measure electrodermal activity is very affordable and non-invasive (figure 7.1): It's simply attaching electrodes to the finger. However, the method is much less precise in identifying the specific emotions experienced (e.g., distinguishing between fear and excitement).

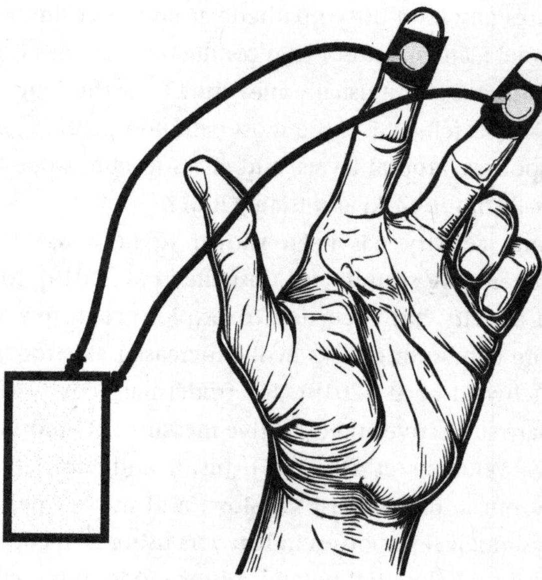

Figure 7.1
Electrodermal activity measures skin response with simple finger sensors.

Combining electrodermal activity with other measures may be necessary for a more nuanced assessment of emotions. Electrodermal activity can also be influenced by non-emotional factors, such as temperature, humidity, and physical activity, which can introduce noise into the data and affect the accuracy of measurements. There is also considerable inter-individual variability in electrodermal activity responses, making it challenging to establish universal baseline values or thresholds for emotional states. We can say that people's responses increased, but the degree of increase can vary considerably between participants. Analyzing electrodermal activity data can therefore be complex, requiring expertise in signal processing and interpretation. Researchers may need specialized software and training to make meaningful sense of the data.

Along with conductance, skin *temperature* changes can be indicative of emotional arousal, cognitive load, stress, or comfort levels, and can vary with emotional states. Researchers can use skin temperature measurements to assess how users' emotional responses fluctuate during interactions with products. Skin temperature measurement has been used in UX testing (e.g., Niu et al., 2019) and music research (e.g., Hu et al., 2022).

7.1.1 Example of Skin Conductance Used in Sound Research

Fortin et al. (2019) used skin conductance responses to detect whether smartphone notifications had been perceived by users. Traditional smartphone notification systems lack the capability to determine if a notification has been noticed without user interaction, leading to potential issues such as redundant notifications or missed messages. The research confirmed that both vibrotactile and auditory notifications triggered skin responses, which differed from responses to arbitrary stimuli. This finding suggests that skin conductance can be utilized to predict the perception of notifications using wearable sensors, potentially improving the efficiency and user experience of notification systems.

7.2 Measuring the Brain: Electroencephalography and fNIRS

There are several ways to measure the brain's response to a stimulus, which can indicate both emotional and cognitive processes. Recent advances in artificial intelligence applied to brain imaging are teaching us about these brain processes. Many of these methods are outside the scope of this book as they involve considerable expense and are not commonly used in UX research. The more common methods for UX or sound and music research include using an electroencephalography and functional near-infrared spectroscopy (fNIRS).

An electroencephalogram (EEG) is a result of the measurement of the electrical activity of the brain by recording from electrodes placed on the scalp. This electrical activity can be categorized into different frequency bands, including delta, theta, alpha, beta, and gamma waves. Each frequency band is associated with specific brain states and functions. For example, alpha waves are associated with relaxation, while beta waves are linked to active thinking and alertness. Moreover, EEG signals can also indicate different brain regions being activated, which can be relevant and indicate which mental processes are activated (e.g., prefrontal cortex with conscious processing, temporal cortex with auditory processing).

There are two main approaches to EEG : those detecting brainwave oscillations and those detecting event-related potentials (ERPs), which are responses to specific events or stimuli. Measuring the timing and location of ERPs helps researchers understand cognitive processes, such as attention, memory, and language, and can indicate if the stimulus is as expected or surprising (Müllensiefen, 2021). With the rise of interest in brain–computer interaction, there are increasingly simplified EEG kits that have brought affordability and use down to the consumer level.

Researchers have used EEG to assess the attention, cognitive workload, and emotional responses of users while interacting with applications, providing insights into how design choices impact the user's experience. High cognitive load may indicate usability issues, while low cognitive load suggests a smoother and more intuitive user experience. EEG can also detect changes in users' attention and focus as they interact with a product or interface. Researchers can identify moments of distraction, engagement, or information overload (Fairclough and Venables, 2006). EEG can also be used to gauge users' emotional responses to advertisements, websites, or multimedia content (e.g., Kwon and Lee, 2020) and product choice (e.g., Lin et al., 2018). Some of these studies have incorporated sound, showing that the inclusion of sound increases attention (Kwon et al., 2022). Studies have also used EEG in testing emotional response to music (e.g., Plourde-Kelly et al., 2021).

EEG is considered safe and well-tolerated by participants, making it suitable for UX studies. However, EEG has limited spatial resolution, making it challenging to precisely pinpoint the brain regions involved in specific cognitive processes. It offers a more generalized view of brain activity. Analyzing EEG data can be complex and typically requires expertise in signal processing and neuroscience. The data can be sensitive to artifacts such as eye movements, muscle activity, and environmental noise. Careful preprocessing and artifact correction are required to obtain clean data.

7.2.1 Functional Near-Infrared Spectroscopy

Functional near-infrared spectroscopy (fNIRS) is another way to measure brain activity. fNIRS is a recent, noninvasive neuroimaging technique used to measure changes in the concentration of oxygenated and deoxygenated hemoglobin in the brain. fNIRS involves placing sensors on the scalp, providing insights into brain activity by highlighting which regions are being activated. Brain activity patterns can reveal the level of engagement and emotional processing associated with different types of media.

Researchers have used fNIRS to assess how users respond to multimedia content, such as advertisements or interactive presentations (Parasuraman and Jiang, 2012). fNIRS has also been applied in gaming UX research to examine the cognitive and emotional responses of players during gameplay. fNIRS has also been used to study our brain's response to music and sound, which is an exciting area, since interference from the sound of the machine had always been an issue in magnetic resonance imaging (MRI) studies. For instance, Jeong and colleagues (2018) used fNIRS to explore how cognitive load changed when listening to different melodic contours. Zhang et al. (2023) used fNIRS to explore how it might improve user experience of bare hand interaction in virtual reality under different sound conditions.

fNIRS is less sensitive to motion artifacts than EEG, making it more suitable for UX studies that involve physical interaction with products or interfaces. fNIRS kits are available at reasonable prices (less than $2,000), and do-it-yourself kits are also available (Tsow et al., 2021). There are also kits that can combine fNIRS and electroencephalography (e.g., Artinis's Starstim), as well as open source, DIY fNIRS kits. However, interpretation of fNIRS data can be challenging, and results may not always be straightforward, requiring expertise to interpret.

7.2.2 Examples of EEG and fNIRS in Sound Studies

Lee et al. (2014) tested responses to sonic notifications during tasks using EEG. They describe the advantage of being able to measure response to passively hearing sounds—something for which none of the subjective methods can account. Audio notifications are crucial for alerting users of electronic devices, but their effectiveness can be compromised in noisy environments or when users are distracted. Traditional methods relying on behavioral performance have yielded inconsistent results due to subjective and environmental variations. To address this, the authors propose an EEG-based method that objectively measures auditory perceptual responses, specifically, mismatch negativity and attention shifting, to assess the usability of audio notifications. This approach was tested in realistic scenarios where users performed tasks amid ambient noise, offering a new perspective on designing effective audio notifications. They showed how ambient background noises can interfere with response to sounds.

Jeong et al. (2018) investigated how cognitive load changes during music listening can inform the design of earcons and auditory icons used in public environments. The research involved sixteen college students who performed melodic contour identification tasks while their behavioral responses (reaction time and accuracy) and cerebral hemodynamic responses were measured using fNIRS. The findings revealed a gradual increase in cognitive load, evidenced by both behavioral data and cerebral hemodynamic responses, particularly in the frontopolar area, where oxygenated hemoglobin levels increased significantly, while deoxygenated hemoglobin levels decreased. The study concluded that understanding the cognitive load threshold and adjusting the relationship between timbre and pitch contour can reduce perceived cognitive load, offering an effective strategy for designing earcons in public environments.

7.3 Measuring the Heart: Electrocardiography, Photoplethysmography, and Blood Pressure

Measurements of our cardiovascular system, including blood pressure and heart rate, can also indicate our responses to stimuli. Blood pressure is rarely used in direct user experience research, although it has been used in some music-based research (e.g., Thoma et al., 2013). More commonly, heart rate and heart rate variability are used for UX and/or music testing, if any cardiovascular measure is to be used.

Heart rate measures the number of heart beats per minute, while heart rate variability (HRV) assesses the variation in time between successive heart beats. Both metrics can indicate a user's physiological arousal, stress levels, cognitive effort, physical effort, and emotional state during UX tasks or when listening to music (Koelsch and Jäncke, 2015). For example, higher HRV is associated with relaxation, while lower HRV may indicate stress or emotional distress. Studies have also shown that different emotions are associated with specific patterns of HRV. For example, increased HRV might be associated with positive emotions like joy, while reduced HRV is linked to negative emotions such as anxiety (Kreibig, 2010). Heart rate variability is therefore a better measure than just heart rate to study cognitive processing. UX researchers use HRV to evaluate whether certain design elements or tasks induce stress or mental strain, which can inform interface improvements. Researchers have used HRV to measure the emotional engagement of gamers while playing video games, for example (Nacke et al., 2011). HRV has been used in music research as well, for instance, in exploring how music impacts the stress response (Thoma et al., 2013).

There are several ways to measure heart rate and heart rate variability. Blood volume pressure (BVP) is a method of detecting heart beats by measuring the volume of blood passing the sensor in either red or infrared light (Diroll and Hlebovy, 2003). Photoplethysmography (PPG) is a recent and popular way to measure heart rate and HRV. It has been used, for instance, to assess stress levels of people when using interfaces (Barreto et al., 2007). When the heart beats, blood is pumped into the arteries, causing the volume of blood in the arteries to increase and decrease with each heart beat: This results in variations in the amount of light absorbed or reflected by the tissue, which can be measured. Although PPG data has not been used much in user experience research, it has been used in conjunction with other measures to explore emotional response (Zhang et al., 2019). PPG is noninvasive and does not require electrodes or physical contact with the skin, making it comfortable for users. PPG sensors are small and lightweight, making them suitable for wearable devices like fitness trackers and smartwatches. PPG data can also be collected using a smartphone camera and specialized apps with a few simple sensors.

Until the arrival of PPG, the most common way to measure heart rate and heart rate variability was ECG. ECG provides insights into changes in heart rate and heart rate variability. When it comes to music, one study used ECG to examine the effects of tempo and mode (major vs. minor) on participants' emotional and physiological states (Khalfa et al., 2008). Another study used ECG to assess emotional responses to familiar and unfamiliar music, shedding light on how heart rate and physiological measures correlate with music familiarity (Dousty et al., 2011).

ECG requires the attachment of electrodes to the user's skin, which can be perceived as intrusive and may affect the user's comfort and natural behavior, potentially impacting the results. Setting up and operating ECG equipment can be complex, requiring expertise in electrode placement, signal acquisition, and data analysis. Specialized software and training are also necessary for data analysis.

7.3.1 Examples of Uses of Cardiovascular Measures to Evaluate Sound Design

Date et al. (2019) discuss the use of heart rate variability in evaluating automobile sounds, and describe their own modifications to retrieve better data from users. The authors propose a method using HRV analysis to assess the feeling of excitement in drivers, which is less reliant on subjective preferences compared with traditional evaluation methods like the semantic differential method. The study finds that changes in the low frequency/high frequency components of HRV correlate with the perceived excitement of different engine sounds, suggesting that HRV analysis can be a valuable tool for sound design in automotive contexts. This research aligns with other studies that explore the relationship between sound and HRV. For instance, the effect of different sounds on HRV has been studied, showing that certain types of music and noise can influence HRV measures, which are indicative of autonomic nervous system activity (Belova et al., 2019).

Pradhan et al. (2022) used ECG to explore the impact of sound in horror movie clips, using AI techniques (wavelet packet decomposition and machine learning classifiers) to evaluate the data. The research addresses how horror sounds affect electrocardiac activity, a topic less explored compared with other music genres. The study uses wavelet coefficient-based features extracted from ECG signals to detect changes induced by horror sounds. These features were statistically compared between two data groups, revealing significant changes after exposure to horror sounds. Nine machine learning models, including naïve Bayes, fast large margin, and logistic regression, were used to classify these changes. The models' performances were evaluated using metrics such as accuracy, precision, F-measure, sensitivity, specificity, and area under the curve. Among these, the gradient boosted tree model emerged as the most effective in distinguishing horror sound–induced changes in ECG signals.

7.4 Measuring the Eyes: Eye Tracking and Pupillometry

Eye tracking is a widely used and valuable measurement technique in UX research. Eye tracking technology typically uses a device (desk-mounted or wearable glasses) that can capture and record eye movements and gaze points while users interact with a product or interface. Eye trackers collect data on where a user is looking, the duration of fixations (periods of steady gaze), saccades (rapid eye movements between fixations), and pupil dilation. The data are used to create gaze plots and heatmaps (visual representations of where users focus their attention on a screen or interface). It can help us to understand visual attention, fixation durations, number of fixations, and scan paths, which can reveal which elements draw users' attention and how they navigate content.

Eye tracking is primarily a visual measurement technique, and its main application is in visual research to understand how individuals visually interact with and process information in interfaces. However, eye tracking can indirectly inform sound-related research in some contexts where visual stimuli are linked to sound-related elements.

Beyond traditional eye gaze tracking, modern systems incorporate pupillometry to measure changes in pupil size in response to stimuli. Pupil dilation can indicate cognitive load, interest, and emotional responses to different elements in a user interface. One study used pupillometry and other measures to assess user engagement with video games (Mandryk et al., 2006). In terms of music and sound, one study used pupillometry to examine how pupil dilation reflects the time course of emotion recognition in human vocalizations, including music (Oliva and Anikin, 2018). A review article highlights the potential of pupillometry to provide insights into preconscious emotional responses to music (Laeng et al., 2012).

7.5 Measuring the Face: Electromyography and Facial Expression Analysis

Electromyography (EMG) measures muscle activity using electrodes placed on specific muscles, usually of the face (sometimes called facial EMG, or fEMG). Specific facial muscles that are typically evaluated include the corrugator supercilia (eyebrows), orbicularis oculi (eyelids), and the zygomaticus major (the muscle associated with smiling). Muscle changes can be detected even when not visible to a human, making EMG more sensitive than automated facial expression analysis (Bell et al., 2018). EMG is effective at capturing variance in the valence of our emotional response (valence refers to the intrinsic positive or negative degree of an emotion; for instance, pleasant or rewarding emotions have a positive valence).

EMG can reveal users' levels of frustration or effort during tasks, such as clicking a mouse or navigating a website. More intense emotions often result in more pronounced changes in muscle activity (Schwartz et al., 1974). Calming and soothing music has been associated with reduced muscle tension and lower EMG activity (Thoma et al., 2013). Upbeat and stimulating music may increase muscle activity temporarily, particularly in response to rhythmic or energetic music. However, the overall influence of music on muscle tension depends on factors such as musical genre and individual preferences (Abraham and Drory, 2014). Bradley and Lang (2000) found that facial EMG activity was significantly higher for unpleasant sounds. Some usability studies have incorporated EMG to assess the emotional engagement and cognitive load of users while interacting with websites, software, or products (Taffese, 2017).

Setting up EMG equipment and preparing participants for data collection can be complex and time-consuming, and researchers require specialized training. And while EMG provides information about muscle activity, it does not directly reveal cognitive processes or emotional states. Researchers may need to combine EMG with other measures to gain a more comprehensive understanding of user experiences.

Another way of measuring the facial muscles to gauge reaction is automated facial expression analysis (AFEA), which uses video- or image-based technology to detect, analyze, and classify users' expressions. It can reveal a few basic emotions such as happiness, frustration, or confusion and assess users' emotional responses to different aspects of the UX (Samara et al., 2019; Salmi et al., 2023). Manual coding is time-consuming, however, but automated techniques have been used more recently, even remotely, for instance, to study facial expression in response to online advertisements (Teixeira et al., 2014). Automated facial coding is possible using cameras in phones, tablets, or webcams, and is therefore now more easily accessible. There are a number of software packages today that offer automated facial expression recognition and analysis. Noldus, for instance, promises between 96 and 100 percent accuracy of detection of happiness, sadness, anger, surprise, fear, and disgust.

While facial expressions can reveal emotions, they do not provide context or reasons behind users' emotional responses. Researchers need to interpret findings cautiously. Some users may find the presence of cameras and facial expression tracking equipment intrusive, potentially affecting their natural behavior and emotional expressions. For instance, someone might smile from nervousness or in a sarcastic way, which can't be easily differentiated from being happy. Collecting and analyzing facial expression data can be resource-intensive, limiting the sample size in studies. Facial expressions can also vary across cultures and individuals, and researchers should consider cultural differences and individual variations when interpreting findings.

7.5.1 Example of Sound Design Study Using Electromyography

James Robb and colleagues (2017) used both galvanic skin response and EMG to explore responses to sound design in video games. They developed a first person shooter game where the player had to collect six objects without taking so much damage from enemy fire that their health level ran out; the health level was therefore a key part of the gameplay that players needed to be aware of while playing, and was also displayed on-screen. The meters were played with and without sound indicating their status. Players also took subjective response surveys and were interviewed after playing. Comparing EMG and skin conductance results between no-sound and sound versions of the game, the results were approximately the same between the two groups, indicating that the sound of health meters did not impact the physiological response of players. While physiological measures did not confirm the importance of the sounds, players themselves reported that they were reliant on the sound warning to let them know about health meters running out, and using the surveys and interviews was key to that understanding.

7.6 Less Common Approaches

In addition to the more common measures above, there are also other less common approaches that may become more relevant as costs decline or the tools improve to become more user-friendly.

The eye blink startle response, or startle reflex, has been used in user experience research particularly in the context of assessing emotional and physiological responses. The startle reflex could also be used to gauge emotional engagement and surprise during user interactions. Walla et al. (2011) applied the test to brand names, showing that least-liked brand names evoked a greater eye blink response. The startle reflex test has also shown how music elicits emotional responses (Bradley and Lang, 2000). However, the startle eye blink reflex is typically associated with unexpected and potentially threatening stimuli rather than music, which is usually considered a non-threatening auditory experience. In UX research, the startle response could be used to assess the effectiveness and user-friendliness of sound design in products. For example, sounds that are too startling or jarring might be undesirable in a user interface, while sounds that elicit an appropriate level of alertness without causing discomfort could be more suitable.

Reaction time has also been used in psychology for over a hundred years, and if you've ever taken an implicit-association test, you've had your reaction time tested. This test was adapted by Müllensiefen et al. (2021) to explore the effectiveness of music in advertising.

Chills, also known as goosebumps or piloerection, are a physiological response that occurs when the small muscles at the base of hair follicles contract, causing the hair to stand on end, and can be measured as a response to sound or music. Chills are often triggered by intense emotional experiences such as fear, excitement, awe, or moments of great emotional significance, like listening to a powerful piece of music or watching a moving film scene. Chills in response to music can vary among individuals, and musical preference, personality traits, and prior musical experiences play a role in the likelihood and intensity of experiencing chills (McPhetres and Zickfeld, 2022). Measuring chills may be useful if the goal is to elicit a particularly strong emotional response.

There are also more invasive measures that are not typically used in UX, including, for instance, measurements of cortisol and endorphin release (Clausen et al., 2023). Cortisol can either increase or decrease in response to music, depending on the type and emotional content of the music. Relaxing music may lower cortisol levels, while intense or unpleasant music may raise them (Khalfa et al., 2003). Researchers have also explored dopamine as it relates to reward and gamification of products (Hamari et al., 2014). Along with dopamine, the endogenous opioid system probably plays some role in pleasure induced by music or products. Oxytocin, often called the "love hormone," can be released in response to music, especially when listening to music that elicits feelings of connection, love, or empathy (Nilsson, 2009). Certain types of music, particularly music that fosters feelings of connection and empathy, can lead to the release of oxytocin (Koshimori, 2019). These measures could indicate a particularly strong response to sound, but as far as I'm aware have not been used to date.

There are also methods that involve very expensive machinery, including fMRI, magnetoencephalography (MEG), and positron emission tomography (PET). These are not used in traditional user experience research due to practical limitations and the nature of UX studies. However, there are emerging areas in neuromarketing and neuro-UX, where these devices have been applied to investigate aspects related to human–computer interaction and user experience. Researchers have used fMRI to examine the brain's emotional responses to music, showing changes in various brain regions, especially the amygdala (Koshimori, 2018). MEG has been used in music research to investigate the neural processes underlying music perception, cognition, and production (e.g., Brattico et al., 2009).

7.7 Conclusions

Physiological measures provide us with valuable insights into the subconscious and automatic responses of users, offering a depth of understanding that traditional self-report methods might not capture. By integrating these physiological metrics with conventional UX methods, such as surveys and interviews, researchers can gain a holistic view of the user's experience with sound, identifying moments of stress, excitement, or frustration that might otherwise go unnoticed.

Our understanding of biological processes and, in particular, areas of brain research is growing rapidly, particularly with the progression of artificial intelligence. Biosensors are becoming much more affordable and ubiquitous, as well as shrinking in size, and it will be interesting to see new consumer-level devices that may be useful in gauging user responses.

The use of physiological measures in UX research is becoming increasingly accessible due to advancements in wearable technology and data analysis techniques. These innovations enable the collection of high-resolution, continuous data in naturalistic settings, bridging the gap between laboratory studies and real-world applications. As the field continues to evolve, the combination of physiological data with other UX methods will likely lead to more nuanced and accurate assessments of user experiences, ultimately informing the design of more intuitive and emotionally resonant products. Nevertheless, despite recent advances, these devices and their data still require a level of expertise and training to take full advantage of, and are novel to most participants and can therefore throw off results. New, wearable devices combined with AI processing will more likely bring about more efficient and effective ways to measures a user's real-time responses.

8 Test Planning and Setup

Once the research questions have been settled on, as seen in chapter 5, and methods selected, as per chapters 6 and 7 (or elsewhere, based on your specific needs), it's important to carefully consider several factors when planning to test UX sound. You will need to identify how and where and when the data will be collected, decide on the number of participants required and their eligibility, decide on any protocols for the test, and plan the location and execution of the test. Additionally, it's necessary to plan out how the data will be collected, analyzed, and reported. Sometimes if you aren't thinking through the analysis, you can miss essential data that you later regret not collecting, such as specific demographic information. Running a small pilot study for larger studies can often save a lot of headaches and time later.

This chapter will delve into these considerations for planning and conducting a user experience test for sound. Sound testing shares some general concerns with other UX tests, such as potential biases and errors, but it also involves unique challenges not typically encountered in other types of UX testing. This chapter emphasizes these distinct challenges and the importance of addressing them in sound-specific testing.

8.1 Choosing the Right Test

As seen in the previous two chapters, there are a great many methods or types of tests that can be used. Of course, the test selected depends on the research question, the point at which the product is in its development cycle (see below), the budget, the time available (see below), and other factors. General patterns around questions are summarized in table 8.1. Note again

I'm only including methods discussed in chapters 6 and 7. There are other methods that I haven't discussed that are common to general user experience research that are not specific to sound, including standard usability testing, surveys and questionnaires, cognitive walkthroughs, think-alouds, and journey/experience mapping. The following are some generalizations about selecting the right test.

Observational methods: Best for understanding use in context and early prototypes, and for gathering behavioral indicators, nonverbal engagement, and natural responses. Captures what users actually do rather than what they say.

Discrimination testing: Best for determining whether users can tell differences between sounds or versions. These methods directly test perceptual thresholds or distinctiveness, so are useful for measuring improvements or updates, or comparing with competitors.

Free-response methods: Best for questions about meaning, associations, memories, mental imagery, or user interpretations, since users generate their own language, metaphors, or explanations, and there is no biased language influencing results.

Sorting/ranking tasks: Best for preference comparisons and understanding relative appeal, thus ideal for measuring improvements or comparing with competitors. Forces users to prioritize and discriminate when making a judgment across options.

Rating: Best for measuring attitudes, preferences, and perceived qualities (e.g., pleasantness, fun, annoyance, quality), when the attributes to be measured are known. Provides quantifiable data that's easy to compare, visualize, and use statistically.

Descriptive analysis: Best for identifying sensory attributes and developing a shared vocabulary for the sound, or developing attributes to measure in other tests. It structures how users talk about perception, which is useful when creating taxonomies or design guidelines.

Temporal/threshold methods: Best for tracking changes over time, adaptation, or repeated exposure effects. These methods measure how perception or feeling evolves with repetition or over sessions.

Physiological methods: Best for measuring emotional arousal, stress, or affective impact without relying on self-reports. Useful when users may not be able to express feelings accurately, or you want to validate their self-reported emotional responses.

Table 8.1
Suggested Testing Methods Depending on Type of Research Question

Question Focus	Suggested Method(s)
Feelings and affect	Rating, physiological, free-response
Preferences	Sorting/ranking, rating
Meaning and associations	Free-response, descriptive analysis
Behavioral impact	Observational, temporal/threshold
Perceptual distinctiveness	Discrimination testing
Attributes and descriptors	Descriptive analysis
User adaptation over time	Temporal/threshold
Usability and interaction	Observational, rating, discrimination testing

8.1.1 Timing of User Experience Tests

The timing of when you test sound as a component of user experience is often up to the company involved and its own timeline. They may be thinking about developing a new product and want to know what sound can bring, or they may be ready to launch and wanting to validate their choice of sound. Or their product may already be out in the market and they may be dealing with complaints or just wanting to improve their product. The timing of the test influences the needs, the research questions, and the methods. In an ideal world, the ideas, concepts, and product are investigated throughout the design cycle, from prototyping through to post-delivery. Earlier tests tend to be done to understand the competition and the users and their needs. Later tests may focus on refining cognitive and sensory impact, usability issues, user satisfaction, and inclusivity.

8.1.2 Available Time

How much time you have available also influences the choice of test. Ethnographic and observational approaches like diary studies, participatory design, and user interviews can take many months to complete. On the other hand, short surveys of sorting tasks, rating scales, or difference testing can gather data in just a few days. Somewhere in between are cognitive walkthroughs, think alouds, and free-response techniques, which can also take much longer to analyze than quantitative data.

8.2 Selecting Participants

Selecting participants for a UX study is a critical stage that lays the foundation for the study's success. The participants you choose in most cases should be representative of your target audience, as their insights will provide valuable perspectives on your product's usability, effectiveness, and overall appeal. Careful participant selection ensures that the data collected are relevant, accurate, and applicable to real-world scenarios. Consideration should be made for user type, behavior, and demographics; the number of subjects to be recruited; and how and where the recruiting is made. Who are your typical users? Who will be the early adopters? Who are the power (high-knowledge) users? Who are the edge cases, or users who are not typical but may also use your product? Be sure to recruit a range of participants to cover the scope of your audience. High-knowledge users are going to bring different types of questions and issues than low-knowledge users. When it comes specifically to sound, however, high-knowledge does not necessarily mean that they need have a high knowledge about sound, but rather about your product. Be sure to choose a testing method that is appropriate for your selected audience.

8.2.1 Recruiting, Sampling, and Screening

There are many ways to recruit and screen participants, depending on your study goals. Sova and Nielsen (n.d.) have posted a series of tips and tricks online. If the sampling method for recruiting participants is not representative of the actual user base, the findings may not generalize to a broader audience. There are many different sampling methods: Miles and Huberman (1994) list sixteen different approaches to sampling, but the most relevant to our purposes are the following.

- *Convenience sampling* involves selecting participants based on their availability and willingness to participate, such as colleagues and employees of the company. It's quick and cost-effective but may not always be representative. Moreover, employee volunteers to studies may be those with an ax to grind, or may have a particular motivation for signing up.
- *Purposive sampling (judgment sampling)*: Participants are chosen based on specific criteria that are relevant to the research. For example, if studying the usability of a new banking app, researchers might select participants who regularly use mobile banking.

- *Screened sampling*: Potential participants are screened based on certain criteria, such as demographics, behaviors, or user type.
- *Typical case sampling*: Researchers select participants who represent typical users of a product or service.

Think through what the inclusion and exclusion criteria of your study should be. Also, where will the study take place geographically, and how will that influence whom you can recruit? (See below for virtual testing.) Also consider the ethical implications of your participants: Are they vulnerable populations such as children or people with cognitive impairments? What benefits (compensation) are you offering participants? You may want to collect a group of individuals who can participate in multiple tests over time, to gauge response to different versions of a product you're developing, and so you might want a pre-screening survey that can store information about their backgrounds, usage patterns, or other relevant information. You may want to check for device compatibility, verify internet connection speeds if you are running virtual sessions, or perform other checks. Remember that audio will play back differently on every device, and virtual sessions run the risk of sounding quite different from what you anticipated. On the other hand, if you are testing a software interface, you may want a variety of devices and setups to test more broadly.

There are some general guidelines and procedures available for screening virtual participants (e.g., Mok et al., 2023). Online surveys can be prone to bots, especially if you are paying respondents. Various approaches to screening out low-quality respondents have been created. For instance, screener questions or attention checks can be used to track dubious answers. Bad data (low quality responses) can occur for a number of reasons, including poor instructions, respondent inattentiveness, and misrepresentation. Screener *attention check* questions where a user can fail by misrepresentation or inattentiveness can be used to categorize poor respondents. In one study, for instance, users were required to wear headphones, and that was outlined in my study recruitment and in my informed consent instructions. However, at the start of the actual test, I included a question about the audio equipment they were using for the test, and found that about 15 percent of users had ignored the requirement and openly admitted they were not using headphones! It was a quick way for me to screen out those responses. I always start from the assumption that participants don't read any preamble or details and dive right into the body of the text, missing important information.

8.2.2 Demographics and Psychographics

When selecting participants, to start, define the specific characteristics of your ideal participants. Consider demographics such as age, gender, location, education level, and occupation. While these may or may not be relevant in terms of your user base, they may have implications when it comes to testing. These factors should align with your product's intended user base to ensure the study's findings are applicable. If your product caters to a diverse audience, ensure that your participant pool reflects this diversity. When it comes to usability questions, in particular, it's not only recommended to include a diverse pool, but a requirement if you want your product to function well for everyone.

As described earlier, when it comes to sound, it's important to understand how human hearing deteriorates with age. Does your sound design hold up for older users? Will it still be functional for users who are outside your typical user base? Including participants with varying degrees of hearing ability and sensitivity ensures that your sound design accommodates a wide range of users. You may be familiar with the mosquito ringtone for phones. This was a ringtone sound above 15 kHz, which only young people could hear (anyone over about the age of thirty tends to have lost those very high frequencies). Kids could use it in school without their teachers hearing it. You can imagine the uselessness of testing such a product on older participants—unless you're testing whether they can hear it or not!

An audiologist typically only tests people's hearing in a range up to about 8 kHz, so many people don't even know that they've lost higher frequencies, even after a hearing test. It's unlikely you'd ever experience any sounds in the high frequency range—people don't tend to like very high frequencies in general—but it's one example of how users could potentially be excluded from your product without testing on a diverse pool. In one user test I ran, I got such significant differences in responses between younger and older participants that I had to go back to a spectrogram (a visualization of the sound's frequencies) to check that there wasn't something in the high frequency range I had missed that might have accounted for the differences in age groups. Depending on the type of study you are running, it may be necessary to pre-screen participants with a hearing test, such as using the Generalized Listener Selection procedure (Fela et al., 2022).

Sound preferences can also be influenced by age (Bi et al., 2017). If you are using known music (existing music) and wish to test it, it is useful to know the participants' tastes in music, as well as their familiarity with the music, as these will both influence the results. Participants will experience more *groove* (that feeling of wanting to move our body along with the music) if the music is familiar. Being a musician also influences the sense of groove (Senn et al., 2021). The results of the study I mentioned above that showed a marked difference in age in responses was likely due to the fact that older adults have grown up listening to higher fidelity music, and thus were able to hear differences that younger adults were not (Collins and Manji, 2024).

In addition to age-related (or other) hearing loss, consider other demographic factors such as gender and cultural background. Studies have found that demographics such as income, education levels, occupation, marital status, political preferences, personality, and geographical location influence music preferences, and to some extent sound preferences (Mellander et al., 2018). These aspects can influence how individuals perceive and respond to sound. For instance, preferences for certain musical genres or sound effects may vary across different age groups or cultural contexts. By selecting one style of music or sounds over another, you're making a statement about the product and its users. Gender also influences preference in loudness (women generally prefer lower volumes; see Kellaris and Rice, 1993). Studies have shown that, at least when it comes to children, boys and girls related differently to musical expressions of emotion (Boone and Cunningham, 2001). In one study I ran, we found a noticeable difference between the tendency of men and women to anthropomorphize, influencing our testing of sound's impact.

Remember, also, that people who have autism spectrum disorder or who are otherwise neurodivergent can experience sounds quite differently from neurotypical people, and they may have heightened sensitivities. Their needs and preferences in terms of how they communicate them and their preferred methods of communicating with you may be different from what you are expecting; thus, your instructions should be phrased accordingly. Be sure the testing environment is free from any other distracting stimuli (see below). Neurodivergent people may also prefer a very clear agenda and structure for the test, and may have difficulty with the ambiguities of free verbalization methods, for instance.

When it comes to sound, thought should also be directed to *psychographics*. Psychographics refers to the study and classification of people based on psychological characteristics, attitudes, beliefs, values, interests, and lifestyle factors (see, e.g., Walton and Evans 2018). If you need solid data to show stakeholders that including music increases the emotional engagement with your product, selecting people who are more emotionally attuned to music may help your case. Having a pre-screening question that describes chills and asks about emotional experiences could ensure you are recruiting people who are emotionally attuned to sound.

Just because there may be differences in response between ages, genders, class, musical training, or other characteristics doesn't mean that users should be excluded from your study for those reasons. Even if you are aiming for highly engaging emotional experiences, don't *only* test on people who experience chills to music, for example. It's useful in particular to see how different samples may impact different user groups differently, because a happy medium may be found that pleases everyone! But if sound is an important part of your usability in particular, ensure that everyone can hear and decode the sound in terms of understanding what it is communicating to your users. Moreover, we don't always know who our end users will be, and the results may surprise us. Therefore, sticking with a strict representative sample can limit our design creativity and our test results.

8.2.3 Product Experience

For some types of products, it may be useful to assess participants' prior experience with similar products or services that involve sound design. Participants with different levels of familiarity can provide insights into how various user groups might respond to your sound. Do you want them to already have experience with the type of product you're testing? Moreover, consider if you wish to recruit participants who already have experience with your particular product (if it is already being used). For example, if I had a software product that people were struggling to learn, I might bring in my users who are struggling and test the addition of sound to see if it helps resolve some confusion, before rolling out changes to everyone.

Identify participants whose goals align with your product's intended use. For instance, if you're testing sounds for a meditation app, participants interested in mindfulness and relaxation are ideal. On the other hand, the people self-selecting into trying out a new product in this realm may be doing so because they are experiencing a particular stressful time in their lives and may not react favorably to sounds, so be careful in your selection criteria and the potential influence it can have on results. Remember as well that participants for sound UX testing should not only be willing to engage with sound, but ideally also be capable of providing thoughtful feedback on how those sounds enhance or hinder their overall experience. A well-chosen participant pool enhances the validity and relevance of your study's findings, guiding you toward effective sound design decisions.

A company may want to use their own employees for listening tests. This may or may not be an acceptable approach, depending on the questions that are being asked. Employees may be able to be completely objective when it comes to answering questions related to what a sound means to them, or how it makes them feel, but they may also be very familiar with a brand and know that you are looking for specific brand attributes to be named. Note also that the employees may not be as diverse as your users. As such, when and how to use employees as test subjects should be carefully considered.

There is also consideration to be made when using participants in multiple tests over time, and this has the potential for *carryover effects* (when one sound bleeds into another). Whether it's safe to use repeat participants depends on your research questions and methods, and the timespans between testing. For many questions, it will be hard for participants to recall any past exposure to sounds in prior tests, but some participants who are attuned to sound or have expertise in the field may have stronger recall. This is especially the case when sounds go beyond simple beeps and we get into testing music, where recall is probably stronger than for individual sounds.

8.2.4 Sound Expertise

As we've seen, sound comes with its own language, and depending on your research question, you may want participants who are familiar and comfortable with this language, who can provide detailed and meaningful feedback to you. When it comes to discrimination tasks, where there are very subtle differences between samples, trained participants or experts are more reliable. In other cases, though, naïve participants (those with no expertise in assessing sound) can be perfectly fine to use. Consider participants with varying levels of expertise or familiarity with sound design principles (Bech, 1992). This diversity can provide insights into how well your sound is understood by different user groups. In general, it is useful to find participants with a listening experience level appropriate to the evaluation task. For most sonic UX purposes, experts are not needed.

The question of whether to use sound experts has been the subject of several studies, and the answer is that it depends on the context and the sounds being tested. When it comes to audio reproduction quality (e.g., in speakers or headphones) experts tend to agree in their evaluations, whereas non-experts tend to have more variability in their response (Bech, 1992). Some music can even cause different physical responses in experts and non-experts (Proverbio et al., 2015), and as described earlier, the brains of sound experts respond differently to those of non-experts. Sound identification and categorization are also influenced by expertise (Lemaitre et al., 2010). One study into whether experts and novices experience sound design differently when it came to auditory displays (a type of interface) found an 80 to 87 percent match in responses, suggesting similar, although not identical responses (McGregor, 2014). Many studies have concluded that, for the most part, preferences, perceived similarities, basic audio quality ratings, and attribute ratings are similar (but not precisely equal) between experts and naïve listeners (McGregor et al., 2014).

There are tests to measure auditory capabilities for creating listening panels. These are too extensive for listeners who are going to participate in a single study, but if the formation of a panel that will be extensively used is desired, these tests may be worth consulting (Ghani et al., 2005). Panels are preferred for complex evaluation tasks, particularly when it comes to subtle differences in sound fidelity and quality evaluation. Subject training may therefore be necessary when it comes to such tasks. If a method of assessing expertise of your listeners is required, eGauge (Lorho et al., 2010) or the ITU-R standard ITU-R BS.2300–0 (04/2014) can be used to outline the methods for assessor screening.

8.2.5 Number of Participants

The number of participants required for a study can vary based on several factors, including the research goals, the complexity of the product or interface being tested, the diversity of the user population, the research methods being used, the need for statistically valid data, and the available resources. When it comes to straight usability, Nielsen and Landauer (1993) believe that about 80 percent of usability issues will be found with the first five participants, but this number is hotly debated. Other researchers have challenged this number (Woolrych and Cockton, 2001), showing that five users are in fact not enough to detect most issues.

Sometimes, practical constraints like budget, time, and resources might limit the number of participants you can involve in the study. Ultimately, the key is to strike a balance between the depth of insights you need and the resources available. It's also important to remember that UX research is an ongoing process, and you can gather valuable insights from even a modest number of participants at various stages to keep the design on track. If your data have a lot of variability in early results, you might need a larger sample size to ensure that any observed effects are not due to random fluctuations or the particular users you sampled.

While there is no one-size-fits-all answer, if your study aims to gather insights about user behaviors, preferences, and attitudes beyond just identifying usability issues, a larger sample size might be necessary. The methods selected can influence the desired number of users. In detailed qualitative research, even a single user can provide at least *some* useful data. Depending on the study's complexity and goals, you might consider involving at least thirty participants to ensure a comprehensive understanding of the user experience and some statistically reliable data (Macefield, 2009). If you plan to conduct statistical analyses to make generalizable conclusions about a larger population, you'll need a larger sample size to achieve statistical significance. The number of participants needed for statistical significance in a user study depends on several factors, including the type of study, the effect size you're trying to detect, the level of confidence desired, and the variability in the data.

The type of data you're collecting (e.g., qualitative or quantitative) and the methods of analysis you plan to use (e.g., parametric or non-parametric) can impact the required sample size. Qualitative research, such as in-depth interviews or focus groups, typically involves smaller sample sizes (ten to twenty) compared with quantitative studies that require larger numbers for statistical analysis.

8.2.6 Presence of Others During a Listening Test

Most user experience or human–computer interaction research into the presence of an observer/tester shows that observation can impact performance, positively or negatively (e.g., Tang et al., 2024). When it comes to music, the presence of others may influence the results we obtain. A study exploring the presence of chills when listening to music showed greater response in solitary listening than when listening in a group, suggesting music listening is more arousing when listening alone (Egermann et al., 2011). Intuition would tell us that listening socially should be more engaging and exciting. However, one study found the opposite to be the case: Emotion ratings were the same between solo and group experiences. Some studies suggest that social context influences the emotional responses to music, meaning studies conducted in a lab would have quite different results compared with studies in someone's home or workplace (Lamont, 2009).

In my work regarding the impact of presentation methods of sound in affective user tests, we found that when we brought people into a lab, they were more positive in their responses, and they wrote much longer responses compared with our online studies. It's likely that our presence in the room had some impact (Collins et al., 2024). This result suggests that having other participants in a room together may influence the results of any sound study we undertake. Depending on whether the product we are investigating is typically used in the presence of others, including other people in the testing environment, may be relevant.

8.2.7 Emotional Influences

There is a well-known aesthetic-usability effect. In this effect, visual appeal influences the perceived usability of a product: "What is beautiful is usable." Unfortunately, there is no available research at this time into what role sound might play in this effect. I suspect that just as with visually appealing products, sonically appealing products can influence perceived usability. Attitudes toward a product (induced by priming tasks, mood, or reading other opinions) have influenced how sounds were perceived (Västfjäll, 2004). For instance, if I primed my participants with positive greetings ("Welcome—we're so happy you're here!"), setting the tone with "Today, we're going to explore some exciting features that we think you'll love. Let's dive in!" or adding encouragement part way through ("You're doing amazing! Remember, every step you take brings you closer to something great"), these can all influence the responses I will get to any sounds I present. Their background mood, such as being in a happy or sad frame of mind, likewise influences judgment (Västfjäll, 2004). A user coming to a test who is angry or upset is unlikely to feel positive toward your design. Gauging prior emotional state may be relevant to your testing if you are testing emotional or affective aspects of sound.

With sound, we have an added factor of the potential for annoyance and noise sensitivity (noise sensitivity is more stable than mood): People who are noise sensitive are far more likely to evaluate sounds as unpleasant (Job, 1999). There is, for instance, the Weinstein Noise Sensitivity Scale, which can be administered if this is likely to be a factor in your research question (see Zimmer and Ellermeier, 1999). On the other hand, sound itself can evoke strong emotional reactions, and emotional response can impact decision criteria in auditory testing (Gerrards-Hesse et al., 1994). One study explored the impact of auditory-induced emotions on the judgment of other sounds, showing that auditory-induced emotional responses mediated their later ability to judge and categorize other sounds (Bergman et al., 2016).

Is it necessary to take into consideration every factor that may play a role in your users' responses? It's unlikely, and may in fact complicate your study unnecessarily. However, if you are getting strange or unexpected responses, it is worth thinking about the role these factors may play in user response.

8.3 Experimenter and Participant Bias

Everyone has some biases, and being aware of biases means researchers can attempt to mitigate their impact through various means. These biases are not unique to studying sound, and are a part of every user experience study. For instance, the *observer effect*, also known as the Hawthorne effect, suggests that people may alter their behavior simply because they know they're being observed. This observation can affect the authenticity of their interactions with a product. Depending on your research question, you may need to account for this in your research. Participants might also provide responses that they believe are socially acceptable or that they think the researcher wants to hear, rather than responding with their true feelings or experiences (known as *social desirability bias*). This is why it can be helpful to hire someone outside the company to run tests. Studies have also shown that people asked in person or on the phone provide more positive feedback than anonymous surveys provide (Nederhof, 1985). Some participants may also tend to agree with statements or questions, regardless of the content, leading to skewed results: This is called *acquiescence bias*.

If participants have prior knowledge or expectations about the product or the study's purpose, it can influence their behavior or responses (Albert and Dixon, 2003). Likewise, users who have already invested in a product may justify its shortcomings and provide overly positive feedback to validate their purchase. This is related to the *sunk cost fallacy*, in which participants who have invested time, effort, or money into a product might be more likely to continue using it, even if alternatives might be more beneficial. *Confirmation bias* may exist, in that participants typically want to interpret information in a way that confirms their preexisting beliefs or perceptions, leading them to overlook contradictory evidence.

If you are presenting multiple sound samples at a time and asking a user to focus on a single attribute, while changing multiple attributes, the results can be skewed by *dumping bias*, where, for instance, a single negative attribute can influence ratings of other attributes (Hermes et al., 2015). On the other hand, carryover effects like the *halo effect* can exist where positive ratings can carry over to other samples.

As with participants, all researchers also come with their own biases. Having a second researcher analyze your data can help to mitigate this problem where data are qualitative rather than quantitative or where coding of verbal responses is necessary. For instance, I compared my own coding of free verbalization responses with those of two students and two different AI software packages in one study. While there was mostly agreement, there were times when we did not agree. In general, the use of multiple scales or methods can help to increase reliability. Even the instructions given can have an influence on the results: Words or imagery used to describe sounds in advance, such as "rate the roar of waves," rather than "rate the sound," can influence perceptual results (Susini et al., 2009).

In addition to the usual precautions in UX testing, when it comes to sound, to mitigate bias, there are a variety of measures that can be undertaken:

- Be sure to prepare all samples in exactly the same way—that is, normalize them to the same loudness, standardize the file formats, and/or make other similar changes. If files were recorded in different formats or different levels of fidelity, attempt to get samples all re-recorded at the same level.
- Ensure consistency in listening practices and tests as much as feasible. There are tests to determine if the user is using headphones, for instance.

Gather information on what device they are listening on, so you can check to see whether this was an influence later.

- Ensure consistency in playback devices, if relevant.
- Use volume level pre-tests for virtual tests, in an attempt to standardize levels—that is, use a sample file and have the user set it to a comfortable volume.
- Use neutral language and don't describe the sounds yourself.

8.4 Presentation and Protocol Considerations

There are additional considerations beyond selecting participants when it comes to UX studies of sound. This section will answer your questions about where, when, and how you should be running sound-based UX tests, or at least prepare you for thinking through what might complicate a sound-based study, so that you can plan around some potential hurdles.

8.4.1 Sound Only Versus Sound with Image, Video, and/or Product

Design teams tend to test sound in isolation of the products; however, how well a sound fits a product in terms of other attributes of that product can be vital in evaluating a sound (Özcan and van Egmond, 2010). As shown in chapter 3, multiple modalities can have a significant impact on the evaluation and interpretation of a stimulus. Previously, I explored the influence of testing methods on free verbalization responses, comparing sound alone, sound with video on the product, and testing the sound in person with the product. I found that more was said about the product in the live tests, although for the most part, responses were quite similar in terms of what the sound meant to the listener (Collins et al., 2024). That test was answering one specific type of question (sound connotation/mental imagery), but for some other questions, I have no doubt that the results would be quite different across different presentation modes. As shown in chapter 3, section 3.5, "Multimodality and Intersensory Integration," other sensory modalities can impact our perception of sound. Including other modalities adds another variable to your tests, but it is worth testing sound with some kind of product image, whether that is remote in a video or still image, or in person.

8.4.2 Choosing the Testing Environment

As with the selection of participants, the needs of the testing environment will be determined by the questions you seek to answer. For questions relating to audio quality, highly optimized rooms designed for audio testing may be preferred, and there are standards available for these purposes (McMullin et al., 2015). Field testing involves gathering insights about user experiences within their natural contexts, which could be their homes, workplaces, or other real-life settings. This approach provides a more authentic understanding of user behaviors, as observations occur in participants' actual environments, reflecting their real-life experiences and habits. Researchers can observe how users interact within their daily routines and identify factors that impact their behaviors. Users face authentic challenges and distractions that might not be replicated in a laboratory setting. However, it may be difficult to simulate different use scenarios, such as under particularly busy or stressful situations. Organizing and conducting testing in various settings can be logistically challenging and might result in less control over variables.

If your product is often used in noisy surroundings, it's valuable to include participants who can test its audibility and effectiveness in such conditions. Testing in office settings may be necessary for workplace-oriented software. Tardieu et al. (2009) found differences in results between lab tests and real-world conditions. Another study tested vacuum cleaner and kettle sounds on a semantic differential scale for roughness, aggressiveness, brightness, reliability, harmoniousness, cost, power/weakness, pleasantness, softness, sharpness, and acceptability, and all of these judgments were different between the field and the lab tests (Fiebig, 2015).

Labs allow us to ensure uniformity in the way that the material is presented to listeners, which can have an impact on response (Volk et al., 2015). Lab studies can allow accurate measurement, but subjects can be influenced by the environment, such as being in the presence of an experimenter, being uncomfortable in the environment, or other factors.

The acoustic properties of the testing environment, such as its size, shape, materials, and ambient noise levels, can all impact how sounds are perceived: Indeed, the environment chosen can influence ratings of ambiguous or neutral sounds (Forys et al., 2025). A test in a concrete space with smooth walls may be highly reverberant and increase the bass content, whereas a test in a lush, carpeted room with thick fabric walls or acoustic panels can reduce reverberation and create a very dry profile. Ambient noise levels, such as from a busy street nearby, can also color the responses, since someone focusing on sound might be influenced by constant sirens or distracting noises outside. Consider that rooms should be reasonably soundproofed and controlled for too much reverberation. Ensure that there are no unwanted noises in the room in which the test takes place, if appropriate—such as from the HVAC system, the sound of cameras, or other devices being used. A well-controlled acoustic environment can provide accurate feedback on the product's sound design without interference from external noises, and consistency between users.

The volume level of sound being played can influence test results, depending on what is being tested (Wycisk et al., 2023). Acoustic features of different playback devices can likewise alter a user's perceptions. More bass below 100 Hz (the subwoofer range), for instance, leads to a greater sense of groove in music (Stupacher et al., 2016), but can also be perceived as threatening. Todd and Cody identified a sound level of about 90 dB as the preferred level to really capture the groove in rock and dance music (in Wycisk et al., 2023). Note that human hearing is also subject to the Fletcher-Munson curve, or equal loudness curve (Porter and Kiernan, 2022), which means we don't hear lower frequencies as well at lower volumes. Turning up the volume thus can increase the perception of the bass frequencies. However, ninety decibels is excessive for most auditory testing, and is loud enough that over even relatively short periods of time can cause hearing damage! I prefer 70–80 dB (80 dB is the maximum level in the European standards for workplaces; Śliwińska-Kowalska and Zaborowski, 2017). This volume is loud enough to hear the bass parts on most equipment, without being overwhelming. However, the length of the test will influence how loud the volume should be: The longer the test, the lower the volume, within reason. Our ears (as both participants and researchers) can get tired over time, and loud sounds for any significant length of time can feel much more irritating. Of course, if you are not providing users with volume control in the product, you should also test at volumes that will match the intended overall volume of your device.

Control over the stimulus can also influence the valence of the results (Wang and Delgado, 2021), as can the ability of the listener to determine the location of the sound (Samermit et al., 2019). Such effects are known through the *affect misattribution procedure*, which is a paradigm that illustrates the misattribution of affective information based on the contexts in which they were experienced (Forys et al., 2025). Most important to the majority of purposes is consistency across tests. If there are multiple tests to be run over multiple days, selecting the same environment to test in is better than changing rooms or locations.

The physical comfort of participants during testing can affect their overall experience and responsiveness. Uncomfortable seating, temperature extremes, or poor lighting might impact their engagement. Consider how other sensory stimuli in the environment might interact with sound. Visual elements like art or posters on the wall, lighting, and even temperature can influence how participants perceive sound and the overall product experience. Bland, neutral colors are preferred (Viollon et al., 2002). Décor should be minimal but natural, rather than clinical, to reduce anxiety.

8.4.3 Remote Online and/or VR Testing Versus In-Person Testing

Online research allows you to reach a larger and more diverse pool of participants since you can recruit users from different geographical locations. This can lead to a broader range of perspectives. Users are in a comfortable environment (typically their own home or office), and are familiar with the equipment, which can lead to more natural and realistic responses as they interact with the product or sound in their usual context, increasing the validity of some types of tests (Honing and Ladinig, 2008). Remote testing is generally more cost-effective since it eliminates the need for physical space, travel expenses, and equipment setup. It's also easier to recruit participants, leading to larger numbers and thus usually higher statistical power. Online studies tend to scale more easily: If you can conduct the test once, you can conduct it a virtually unlimited number of times, as long as you find the participants. (For a general review of using virtual reality (VR) to test sensory responses and emotion, see Magalhães et al., 2024.)

However, the downsides to online testing for sound are many. Experimenters might miss out on certain contextual cues that can be observed during in-person sessions, such as body language, nonverbal cues, and environmental factors. There is also a greater potential for misunderstanding instructions (this can be reduced by piloting the online tool and using suggestions from users to improve the wording and layout of its design) (Tan et al., 2021). There are higher drop-out rates: It can be that the participants get bored or distracted, or are not motivated to perform well (Zhao et al., 2022). Tests should be shorter than about twenty minutes at most (Bech and Zacharov, 2006). Perhaps most importantly, online studies are open to fraud (bots, etc.) (Zhang et al., 2022).

In music perception testing, studies have shown the validity of remote testing (Peng et al., 2022). When it comes to auditory psychophysics, however, online/remote testing has been mostly avoided (Eerola et al., 2021), since there is so much that is out of the experimenter's control. Users may not understand the impact of background noise on the study, and be subject to background noise interference. For instance, they may be unaware that construction noise beside their office may impact their results when they are actively listening. It is difficult to know the hearing capabilities of the participant, the volume of their playback devices, or the quality of their equipment when using remote testing (Peng et al., 2022). There are measures that can be employed to select participants based on hearing status (Mok et al., 2023). However, for most tasks, hearing screening is not typically completed, as accurate screening requires equipment and expertise (Otto et al., 2001).

Of course, when it comes to physical products, online tests remove the sense of touch and the scale of a product. There is a close perceptual interrelation between touch (haptics) and sound, and the absence of one may well influence the perception of the other. The impact of the lack of touch can depend on the product and questions to answer, of course. In some instances, a simulation can be enough to answer questions, although as far as I'm aware, few studies exist about evaluating sound in a virtual product and comparing that with physical results (Chu and Kao, 2020). In one study I ran, we explored responses to robot sounds, and found little significant differences between virtual and in-person responses; however, we did note that the in-person responses were longer and slightly more positive (Collins et al., 2024). One study of physical versus virtual products (not looking at sound) found that physical models received more positive feedback than virtual models, and that the evaluation methods significantly influenced the result (Voit et al., 2019). Another compared the interactions in a virtual reality (VR) and in a real building, and found few quantitative differences, but significant qualitative differences (Kuliga et al., 2015). It seems clear that some products are better suited to virtual prototyping than others (Voit et al., 2019).

When remote testing, some precautions should be taken, including ensuring that the user will not be exposed to harmful volume levels, that confidentiality issues (potential for leaks, hacks) will be considered, as well as verifying that the user's hardware meets the required stimulus quality. There are testing protocols for internet experiments designed to account for variations in users' playback devices for internet-based experiments, to assist in standardizing volume levels, detecting playback formats, and providing basic equipment tests (Zhao et al., 2022; Wycisk et al., 2023). Some calibration may be necessary in some types of tests, and collecting information about the specific make and model of equipment used by participants may also be beneficial. Different setups and incorrect calibrations between participants can impact test results, depending on the stimuli and the question being researched.

The Acoustical Society of America Technical Committee on Psychological and Physiological Acoustics launched the Task Force on Remote Testing (https://tcppasa.org/remotetesting/) in May 2020 with the goals of surveying approaches and platforms available to support remote testing and identifying challenges and considerations for prospective investigators. The results of this task force survey were made available online in the form of a set of Wiki pages and summarized in their report (Peng et al., 2022). Some recommendations include putting platform, browser, and audio checks at the start of the study, followed by biographical questions, instructions, and then the experiment (Honing and Ladinig, 2008).

8.4.4 Headphone Versus Speaker Tests

The quality of audio playback equipment, such as speakers or headphones, can impact how accurately participants perceive and evaluate the product's sound design. Imagine the difference between movie theater speakers and a cheap television's output. Similarly, the quality of the equipment can greatly impact what we can hear, what we focus on, and how it impacts us. Whether or not your participant has low-quality speakers or is wearing headphones (and what type of headphones) can definitely play a role in quality evaluations, but probably plays a much lesser role when it comes to connotation (mental imagery). When it comes to remote testing, we often do not know the quality of the equipment of the participant, and nor do we know if their equipment is set up and calibrated correctly. Sometimes people will put all of their surround sound speakers on one wall to avoid cables (see Goodwin, 2009)! We then have to consider the additional variable of different equipment that may impact the study.

Headphones can screen out (or at least reduce) external noise sources, minimizing distraction for users, and can easily separate signals to two ears, useful for some types of tests (particularly when binaural audio is involved). They can eliminate reverberation from the room and eliminate some of the problems associated with incorrectly set up loudspeakers. However, headphones tend to be poor at reproducing very low frequencies. Different headphones have different frequency responses (i.e., volume levels of frequencies can vary widely), and the frequency response of the headphones can influence the results of some studies that require accurate identification. Different headphone models have even been shown to modulate brain waves differently (Naal-Ruiz et al., 2021)! As described earlier, there are methods to test whether users are wearing headphones (Milne et al., 2020), and studies have shown that as many as 49 percent of users tested this way failed the test (meaning they were not wearing headphones, when they said they were), so if playback device is a consideration, using such a headphone test is necessary (Woods et al., 2017).

8.4.5 Deceptive Testing

Deceptive testing typically involves deliberately misleading participants about the true purpose of a study to elicit natural behaviors without bias. For example, we might ask them questions about the visual design and get them to focus on that, and then later ask them what they remembered about the sound, or if the sound was distracting. These types of minor deceptions—perhaps *misdirection* is a better word here—should not present significant concerns with regard to the ethics of testing, but as with any deception, deception can be abused and should be used with caution.

8.4.6 Test Duration, Order, and Type of Presentation

The order of sounds played should be randomized where possible to reduce bias created by playback order and/or listener fatigue. Sound may have the *earworm* (known in the scientific literature as involuntary musical imagery) capabilities of sticking around in our head long after the stimulus ends. The influence of one sound may thus bleed into another, in a carryover effect. If the purpose is to evaluate sound quality, or detect subtle differences in sound characteristics, longer intervals between samples may be necessary to prevent auditory masking, adaptation, or carryover effects. Likewise, if a sound sample is quite long, or samples are very similar to each other, longer intervals between sounds may be necessary. The length of interval between sounds being played should be standardized (Reinhardt and Niese, 1970), or the user should have control over playback.

One study found that audio quality preferences are lower when a long stream of audio is played continuously, rather than cutting up the sound into segments that are randomized (Borowiak et al., 2014). The study was specific to audio quality, but what we can take from this is that an individual sound may be received differently from multiple sounds heard in the context of a product used for hours on end. Depending on your research question, you may wish to test individual sounds to see what they communicate to your users. A user test should also consider the uses of the sound for longer lengths of time.

Test duration can matter in a UX test. The appropriate test duration depends on various factors, including the complexity of the product or service being tested, the objectives of the test, the volume levels, the listening environment, the target audience, and the resources available for conducting the test. A longer test duration can sometimes better mimic how users interact with a product in real-world scenarios: It allows participants to explore different features, navigate various sections, and potentially encounter issues that might not surface in a shorter test. However, participants' attention and engagement can wane over longer test sessions (i.e., those over about twenty minutes). Fatigue can lead to inaccurate results as participants might rush through tasks or overlook certain aspects. Fatigue is particularly important when it comes to listening tests, as the attention required to focus on sound can be considerable. *Listener fatigue*, also known as *ear fatigue* or *listening fatigue*, can occur when someone is subjected to repetitive or loud sounds over an extended period of time (Schatz et al., 2012). Individuals experiencing listener fatigue may find it difficult to concentrate on the auditory content, impacting their ability to differentiate between sounds or distinguish fine details. Listener fatigue can also lead to irritability, discomfort, and even stress, impacting UX results.

In general, finding the right balance between test duration and the depth of insights you want to gather is key. It's important to clearly define the objectives of the test and tailor the test duration accordingly. Pilot testing and continuous refinement of your testing approach based on feedback and observations can help you determine the optimal test duration. One study suggests that a maximum of about ninety minutes should be scheduled at one time (Schatz et al., 2012). Other studies suggest twenty to forty-five minutes as the maximum (Lemaitre et al., 2016). Be sure to randomize the order of stimuli when there is a longer test, to account for listener fatigue in your users.

8.4.7 Retrospective Versus Real-Time Testing

When it comes to some types of tests, ratings made retrospectively (after the sound has stopped), particularly when attempting to remember and judge emotional response or sound, can have discrepancies compared with real-time or concurrent ratings. In one study, for instance, researchers found that retrospective ratings significantly overestimated unpleasantness compared with real-time measurements (Bruun and Ahm, 2015). Most studies, however, have adopted a retrospective, post-performance approach, in which ratings are provided after stimulus exposure. There are some approaches and tools available that allow for real-time collection of emotional responses to stimuli (Zentner and Eerola, 2010). The best time to ask questions is often immediately following each task (post-task), rather than waiting until the end of a longer session (post-study), since sound is transient, and our memories are faulty.

8.4.8 Pacing and Timing

Automated or self-paced tests can be used. Self-paced tests means the user has control over when (and usually how many times) a sound is played. Paced tests, on the other hand, control playback timing and ensure consistency across users (and you may prefer not to allow the user to control how many times they hear a sound). If automating the playback, be sure to standardize the interval between sounds, and leave enough time for the user to read and answer any questions. Particularly when using crowdsourcing user-testing websites like Testable Minds, Cloud Research, MTurk, or Prolific, include an attention check in the survey to prevent users from just clicking through quickly. Have a sense of how long the test should take, and remove low outliers who clearly clicked through and did not take at least a set amount of time to complete the test.

8.5 Conclusions

I've introduced many different considerations in planning for sound UX testing. In my experience, the most important are to develop consistency as much as possible for sound tests, and to reduce the extraneous variables as much as possible. Equipment and location vary considerably, and if you are comparing quality between sounds, it can be difficult or impossible for some (particularly untrained) people to hear a difference on low quality speakers. Running online studies, for instance, using crowdsourcing websites can mean most are listening to your sounds on low quality mobile phone speakers. Likewise, consistency between quality levels, and especially volume levels, can make a difference to responses to sounds. If you recorded your competition's sounds at one volume and compare them with your sound at another volume, the volume levels can skew the results.

When selecting participants for sound-related studies, it's important to define specific characteristics that align with your product's intended user base, considering factors like age, gender, location, and occupation, as these demographics can influence how users perceive and respond to sound. Additionally, to ensure inclusivity and obtain accurate data, it's important to include participants with varying degrees of ability and diverse psychographic profiles, as preferences and sensitivities to sound can significantly differ across age groups, cultural contexts, and personal experiences with music.

Consider if it's necessary to establish a panel of regular users and run hearing tests on those users. Be wary that employees can be useful for some sonic user testing, but only to a limited extent for most tests, and bringing in users who are completely unfamiliar with the product or brand is helpful. Test on a variety of users, including those with different levels of hearing ability and musical familiarity.

Remember to align your tests and test protocols with stakeholder goals and your research questions. If subtle differences between sounds need to be determined, be sure to use equipment with which everyone can hear those differences. Use stakeholder feedback to make adjustments to the research questions, test design, participant selection, or sound elements being evaluated. If a product is designed for a particularly loud or quiet environment, be sure to test under those conditions. Document your setup, participant selection criteria, and data collection methods meticulously for your reports, and for future replicability or comparison.

9 Analysis, Interpretation, and Presentation of the Results

As we saw in the previous chapters, there are many kinds of data that we can collect about the user's experience of sound. We can roughly break these down into qualitative data (e.g., user comments, free verbalization, observations, interview transcripts) and quantitative data (numerical data such as task completion rates, rating scale results, and psychophysiological data). Most commonly, we will have a mixed methods approach and have both quantitative and qualitative data. We can also apply some quantitative analysis methods to qualitative data in some circumstances.

The analysis, interpretation, and presentation of results from sound UX studies do not differ from the approaches to other UX data. As such, the reader is encouraged to consult other user experience books in addition to the information I provide below. There is far more to data analysis, statistics and interpretation, and visualization and reporting than can be covered here, but a quick overview will point newcomers in the right directions to follow up more deeply into the areas that are best to answer their research questions.

The first step in analysis is to pre-process all the data we have gathered. This involves cleaning up the data, removing any outliers, and deciding how to deal with any missing data, ambiguous data, or contradictory results. From there, we can employ a variety of analysis techniques. Often, we employ more than one analysis method on the data, potentially through triangulation (using multiple methods or data sources) to validate or confirm findings from one method. In particular, we need to employ methods that are suitable to answering our research questions.

Cleaning the data involves removing duplicate respondents (if anyone went through the test twice). Remove any speeders as well—those who failed attention checks, or went too quickly through the test. Did you undertake the test yourself? If so, how long did your test take? Sound takes time to play back, and I've experienced people swearing they completed my study, but the time taken was less than the sound content alone would have taken to play back! Remove those who answer the same question the same way (e.g., always answers the same on every multiple-choice question, or makes a straight line down a Likert study).

Decide how to handle missing data. For instance, what would you do if they got to question five out of ten but dropped out? My general practice has been to remove incomplete data, but there are times when you may not have enough respondents to make removal feasible, which can complicate the analysis. If questions have been randomized, drop-outs might occur at different question numbers for different questions, so removal is typically easier for the analysis than having to accommodate the differences. This means it's advisable to recruit a few more people than you need, to have back-up.

Look for outliers in the data that may have been skewed for some reason. In one online study, I was interested in how long it took respondents to answer the study, but didn't explicitly ask them to complete the study in one sitting. Most respondents took between three and five minutes, but one was nearly thirty minutes. Clearly, they had stepped away from their desk at some point during the study. If timing is important, be sure to express that to your participants and decide how to handle outliers.

Decide as well how you will deal with low quality responses: Were they lazy in their open-ended responses and gave you only one word when you were expecting several sentences? Decide how to deal with data where the users may not have understood the question. In one test I asked users to respond to sounds and write down any mental imagery that came to mind. Instead, one (clearly expert sound designer) critiqued the use of effects in the sound design. How you handle these types of data is up to you, but generally, my usual practice is just to eliminate any of the obvious outliers. Of course, you don't want to eliminate potentially useful data, so this practice can take some massaging and experience to get right.

Since data analysis can be a book in itself, I will focus in this chapter on some methods I have found useful that may not be covered in existing texts. The rise of artificial intelligence has meant that there are many new tools that can assist us in data analysis, especially when it comes to qualitative data, which is often more complicated to analyze and summarize for stakeholders.

9.1 Qualitative Data Analysis

Qualitative data analysis encompasses various methods and approaches, and is aimed at understanding patterns, themes, and concepts within the data collected. There are a variety of methods, but a few of the most common are content analysis, thematic analysis, and discourse analysis. There is far too much involved to cover here, and I would direct the reader to Udo Kuckartz and Stefan Rädiker's *Qualitative Content Analysis* (2023) for further reading.

9.1.1 Content and Thematic Analysis

At the heart of qualitative data analysis is the coding process. This is the time-consuming task of reading through all the data and assigning codes to text that represent specific ideas or concepts, typically ones that recur regularly. These codes may be assigned in advance if you are looking to answer a specific question (i.e., inductively building the codes). For instance, in exploring the free verbalization responses to robot sounds, I was looking for anthropomorphizing tendencies, and so decided in advance to code all the answers for anthropomorphism (Collins and Johnston, 2023). However, in analyzing the data, I also discovered other themes that recurred, and so then decided to also code for emotion words and onomatopoeia. In a study of user experience, I might create the codes *usability issues* or *onboarding* and tag all phrases relating to this particular aspect of user experience. I may also discover that there are strong emotional connections made, and so also code for those aspects. There are AI tools that will look for semantically similar terms and aid in coding by themes, but in my experience, they are still not as accurate as a human coder.

Content analysis attempts to transform qualitative material into quantitative data using clear and systematic approaches (Krippendorff, 2018). It involves coding the content into categories and then interpreting the meanings or patterns that emerge. For instance, I may be interested in the frequencies of recurring words (including semantically similar terms). The process can be automated or performed manually, but can be very time consuming, particularly when it comes to manual analysis, since there are so many words that mean roughly the same thing in English, and automated analysis can be complicated by opposites. Imagine, for instance, if one participant said a sound made them feel not happy, but another said a sound made them feel sad. If I've automated to look for references to happy, the first might get flagged as happy.

Themes in qualitative data can reveal common needs, desires, and preferences among users. For instance, if a strong theme revolves around the appreciation for personalized content, this could lead to a recommendation to enhance and highlight the personalization features of a product. The process of identifying and analyzing themes is often described as a more intuitive form of analysis. However, as with content analysis, the method involves some subjective decisions, and might drive results that don't necessarily align with reality. As such, it's a useful practice to have more than one coder where possible, either to develop themes and/or to categorize responses into those themes.

There are a variety of new tools that have been developed to automatically code data, including, for instance, Lexalytics, Delve, Atlas.TI, and MaxQDA. However, in my experience, software packages can give quite different results, so the interpretation can be quite different between packages. Automatic coding of the data leaves much to be desired, but can be useful when a second human coder is not available. However, there are still nearly always some subjective decisions to be made. There are also some useful tools to assist in coding some types of response data, such as the Geneva Affect Label Coder, which provides affective terms and their synonyms (Scherer et al., 2013). Many new AI tools also provide some methods to apply quantitative methods to qualitative data.

9.2 Quantitative Methods for Qualitative Data

Today's natural language processing software, and AI-enhanced software in particular, offers us additional approaches and opportunities to explore qualitative data. For instance, such tools can provide sentiment analysis, emotional analysis, and word frequency analysis (including synonyms). These are very useful on larger datasets to quickly get a sense of the data.

9.2.1 Sentiment Analysis

Sentiment analysis is typically used for large-scale text data to gauge an overall response of positive, negative, or neutral sentiment (figure 9.1). For instance, a politician might run a sentiment analysis on social media that uses their name to gauge how constituents feel about them at any particular

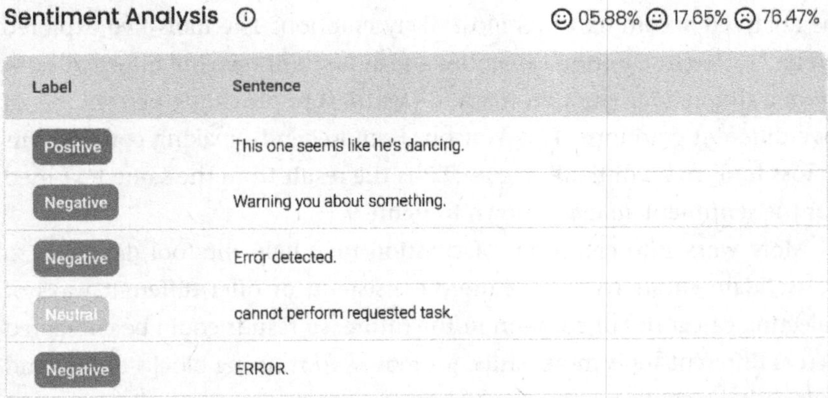

Figure 9.1
SpeakAI sentiment analysis tool. The score (top right) is the overall percentage ratings of sentiment as positive, neutral or negative. The app also gives individual scores for responses.

time. In 2023, I ran a sentiment analysis on a block of text through several different software packages, including SpeakAI, Textrics, and IBM Watson. We also had three humans code sentiment manually, and then compared our results against each other and the software, and for the most part found they were fairly consistent. However, there were definitely some discrepancies. Most sentiment analysis systems do not explain their coding choices or how they break down the analysis. The automated results were at times surprising. For instance, *condescending laugh* was coded as positive, but *sarcastic laugh* was coded as negative. The software coded *wrong* as negative but *incorrect* as neutral. Software also struggled more with ambiguous terms, but these ambiguities are also difficult even for human coders to classify. Our understanding (and therefore bias) of the context of the results likely influenced our interpretation of more ambiguous results (Collins and Johnston, 2023).

9.2.2 Emotion Analysis

Emotion analysis typically goes a bit further than the simplistic positive/negative of sentiment analysis, categorizing text into a series of emotions. How emotions are divided and categorized varies from program to program, however, which can make comparing different analyses difficult. Analysis tools are varied in terms of the number and types of categories they use, but effectively run the same way as sentiment analysis: A large block of text can be analyzed according to its most likely emotion. The tools I've explored all had different emotion categories and systems of relaying information—from a simple four emotion method (Textrics) to providing percentages of five different emotions (IBM Watson, Komprehend), making comparisons across tools more difficult. Figure 9.2 is the result from the same text used for the sentiment analysis shown in figure 9.1.

More work into categories of emotion may help the tool designers to consolidate emotions and be more consistent, or offer different ways of selecting categories of emotion in the future, so results could be compared across different tools more easily. Moreover, in running blocks of text and comparing responses between software, there was significant disagreement between analysis tools, even on responses to sounds deemed unanimously negative or positive in the sentiment analysis we ran. Despite some challenges in accuracy, automatic coding and scoring of large blocks of text into summaries of the emotions work well enough for most user experience purposes. The amount of ambiguous data is generally small, and so will likely not skew overall results too much in any one direction that might result in different design decisions.

9.2.3 Word Frequency Analysis

Perhaps the most simplistic approach to quantifying the qualitative data is to use word frequency analysis. Complications arise with synonyms or near-synonyms and deciding at what point to clump them together into one concept. Often, the text still needs to be manually stemmed or lemmatized (grouping inflected forms of a word, such as *improve, improvements, improving*). Word frequency clouds can be created, and such ways of visualizing the data can be useful to quickly summarize and display the key themes that arise in a large body of responses (figure 9.3).

9.3 Statistical Methods

The statistical method applied depends, of course, on the type of data gathered, as well as the purpose of the test, or the questions being asked of the data. Moreover, the number of samples (and whether they are independent or dependent), number of participants, whether there is a comparison being made, and other factors will all influence the method of analysis.

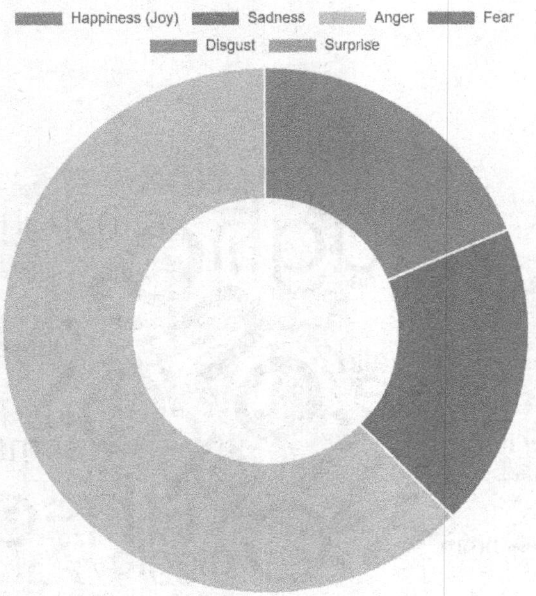

Total Feedback Analyzed	16
Happiness (Joy)	3 (%18.75)
Sadness	3 (%18.75)
Anger	10 (%62.5)
Fear	0 (%0)
Disgust	0 (%0)
Surprise	0 (%0)

Figure 9.2
Sentiment analysis in Benty AI.

Figure 9.3
Word cloud of some responses to interface sounds, generated in Wordclouds.com.

The type of data is important to the method selected: When you're working with data, there are different types depending on how the data are measured. Some data just put things into groups or categories—such as if something is correct or incorrect, or sorted into names like *sharp* or *sound one*—this is *nominal* data. If you're putting things in order, such as ranking your favorite sounds from best to worst, that's *ordinal* data. Then you have *interval* and *ratio* data—these are more math-friendly, and lend themselves to calculations like averages, standard deviations, and correlations.

Different methods are also selected depending on whether or not an individual sample is characterized, or if two or more are compared with each other. In the case of several samples, testing generally involves *dependent* samples; that is, the data originate from the same panel or the same tester. The same testers have, for example, examined both a sample of the standard product and a sample with the new, improved sound. *Independent* samples mean that samples are tested by different panels, for example, when a user test is conducted in parallel in different towns or at different times.

A brief guide to statistical methods is included here; however, to fully do the subject justice would require at least another book.

9.3.1 Difference/Discrimination Tests

Goal: To find out if there's a difference between two sounds. These tests help you figure out if the change in sound made a meaningful difference in an area of user response, like comfort, speed, or satisfaction.

Example: You change the interface sound of a button click and want to know if users actually notice the difference.

Statistics: If the same users test both sounds, use a paired-samples t-test (parametric) or Wilcoxon signed-rank test (non-parametric). If different users test each sound (e.g., two groups of participants), use an independent-samples t-test or a Mann-Whitney U test. If you have more than two versions, use an ANOVA (parametric) or Kruskal-Wallis test (non-parametric).

9.3.2 Sorting and Ranking Tests

Goal: To find out if users have a consistent preference for one sound over another.

Example: You play for the users three notification sounds and ask them to rank them from most to least pleasant.

Statistics: Use the Friedman test for dependent samples (assuming the same group of people ranked all sounds). For independent samples, use the Kruskal-Wallis test.

9.3.3 Rating Scale-Based Tests

Goal: To measure how strong a certain attribute is across sounds, for example, how annoying or pleasant a sound feels, using a rating scale.

Example: Users rate each sound on a scale from 1 (not annoying at all) to 10 (extremely annoying).

Statistics: You can calculate means and standard deviations to describe your results. If you're comparing groups, you can use t-tests or ANOVA, but only if your data look normally distributed. If your data are skewed (e.g., most people give scores at the top or bottom of the scale), switch to non-parametric tests like Wilcoxon, Friedman, or Kruskal-Wallis. Note: Just because the scale looks as if it has equal steps doesn't mean users actually treat it that way. People tend to avoid using very high or very low numbers, which can affect your data's normality.

9.4 Reporting Results

The reporting of sound-based user experience studies should follow the standard protocols for reporting on UX research. Knowing the audience of the report and understanding their knowledge about sound are critical to successfully communicating about sound. Unlike a visual design aspect that can be included in a printed report, the sound cannot be communicated without adding or embedding playback of files. It's possible, for instance, to build a slide deck or document with the sounds embedded so that people reading the reports can hear what sounds were tested. Defining terminology may be necessary as well. For instance, if you were specifically testing timbre of a rhythmic motif that may be used in an interface, you may have to define timbre for the audience with examples that they can hear.

Consider separate reports for executives (who need the main points and summaries) and the design team (who may need more details on what was or was not working in the sounds under research). Hopefully the design team will know more about sound and be able to understand some nuances and terminology that the average person may not know, but this should not be taken for granted. Be sure to include details about how the tests were conducted, who the participants were, and under what circumstances the sounds were played to those participants, since these details can be a point of contention, especially when stakeholders feel invested in a style or genre of music that *they* like, even if the results show that their users do not like it.

Educate stakeholders on any terminology that they may need to know for the presentation, ideally with sound samples so that they clearly understand whatever attributes you were testing. Given that most audiences are not familiar or may be uncomfortable with sound terminology, consider helping them to express their points of view with alternate terms. For instance, instead of saying *timbre* you might talk about the *texture* of the sound. They may be more comfortable talking about *volume* than *amplitude*, and *pitch* than *frequency*. Be sure to explain any sound-specific terms you are using.

Avoid over-interpreting non-significant results: Consider effect sizes and trends and whether those should also be reported. Consider also including the ecological validity: Was the sound tested in a realistic setting? Was the sound paired with the actual task or product it was designed for? Where feasible, triangulate statistical data with qualitative data: Add quotes or open-ended feedback received to give depth to numeric results so you don't overwhelm with numbers. Think about how best to present this qualitative feedback, such as with word clouds or annotated screenshots.

If you are presenting the material in a demonstration, consider setting aside time for a listening demonstration to illustrate your points. Understanding who the audience is will be key to a successful listening demonstration, as different stakeholders will have very different needs from the presentation. Are you demonstrating to the creative team that includes sound designers or composers, or are you demonstrating to executives who may have little interest in the nuances of sound? Consider the goals of your presentation. If you are arguing for higher quality speakers in a product, then demonstrating the difference that can clearly be heard by most people will help to illustrate your results.

Generally speaking, you want high quality hardware to demonstrate any sound examples. This may mean bringing in external speakers yourself for the presentation, rather than relying on, for instance, a television's built-in speakers of unknown quality (or, worse, no audio at all in the room!). Higher fidelity audio can really demonstrate the effect you want to express, so ideally use the best speakers and highest quality sound files you can.

If you have the opportunity, consider the room in which your presentation will take place (i.e., where the sounds will be played back). The room can color the demonstration, and make sounds easier or more difficult to hear. Is there a loud HVAC system that might kick in during the demonstration? Can you control for that? Is there a lot of traffic noise from the street, and if so, can you use a different room? Are there a lot of reflective surfaces in the room, and if so, can you hang a blanket, draw the curtains, or otherwise impact those surfaces? Poor acoustics can distort the sound. There is also some evidence that people hear better in darkened rooms—when visual distractions are limited, it can help people to focus on the sound.

What will be the distance from the audience to the speakers, and what is a comfortable volume level at which to present the sound? Ideally, bring along a colleague to test the sound with you in advance. The further from the sound source, the more the background noise will potentially interfere with what is heard. For instance, when it comes to speech, people with full hearing range need a speaker's voice to be at least 6 dB louder than any background noise to hear clearly. But someone with hearing loss may need a 16 dB difference. Tolerance for noise also decreases with age and with neurodiversity. If your audience is older, consider a slightly louder volume than what you would play back for younger audiences if there is a lot of background noise. Or you may consider asking the audience at the start to tell you what a comfortable level for them is. Once you have set the volume level, be sure to be consistent across all the sound files that you may be playing—normalize them to the same volume level in advance.

Be sure not to talk over the sound, particularly while playing longer sound files. It's tempting to treat the sound (especially music) as background noise and expect a listener to listen to both you as the speaker and the sound files, but it's not possible to focus on both! Consider also visualizing sounds if there is a clear visual difference: envelopes, spectrograms, or waveforms with timings may be beneficial.

If you are presenting virtually, be sure to understand the settings in the virtual meeting tool you are using. Zoom, for instance, offers several different features that can be beneficial when speaking, but can interfere with presenting audio. I was once showing a friend a new guitar on Zoom, but he couldn't hear any of it, since Zoom had decided that it was non-speech sound and filtered it out. There are settings to adjust to accommodate these other sounds. Be sure to also have the audience turn on these settings, and turn on high-fidelity mode, in stereo. They may also need to turn off noise suppression. In Microsoft Teams, noise suppression is turned on by default, for instance. Be sure to test your software with a colleague, and write up (or make a video of) simple instructions for your audience to change their settings, and do this right at the start of, or in advance of, the presentation. If high quality audio is going to be necessary, and you have the opportunity, you may consider sending the audience some speakers or headphones, or having someone visit to set up some external audio for them in advance.

Audiences may want to hear sounds more than once, and may want to hear them on their own time, outside the meeting. If possible, set up a website where they can test out the sounds on different devices or at different times and volumes that are suitable for them. This way, they are hearing the sound on equipment that they are familiar and comfortable with, and can hear the sound under different conditions that their own users may experience.

9.5 Conclusions and Take-Aways

The analysis of sound-related data isn't unique compared with other types of user experience research. Nevertheless, it's worth considering that sound can introduce many variables into a study that may need to be accounted for: The equipment, the room, the volume level, and other factors can all influence the results. If you are collecting the data, running additional inferential statistical analyses such as regression analysis, correlation analysis, non-parametric tests, or ANOVAs can help to determine if such variables had an influence on your data. For example, suppose you collect data on user satisfaction with sound in different conditions:

Speaker type (three levels): Built-in laptop speakers, external speakers, headphones

Volume level (two levels): Low, high

Environment (two levels): Quiet, noisy.

You could set up a three-way ANOVA with user satisfaction scores as the dependent variable, and speaker type, volume level, and environment as the independent variables. This setup would allow you to see the main effects of each variable (e.g., does speaker type affect satisfaction?), along with interaction effects (e.g., does the effect of speaker type on satisfaction depend on the volume level?). If significant interactions are found, post-hoc tests can help to pinpoint where the differences lie. Regression analysis can further highlight how much variance in satisfaction scores is explained by the combination of these variables and their interactions.

There are some special considerations when it comes to reporting and demonstrating the results of sound to stakeholders, as outlined here. Ensuring you have optimal presentation setups is key to driving home key points about sound quality in particular. If your stakeholders can't hear the difference between two sound files because they're hearing the examples on poor quality desktop or laptop speakers, they may not buy your argument for increasing a budget for sound.

10 Conclusions and a Final Word

Sound is not just an accessory to user experience; it is a fundamental component that can transform the way users interact with products. As we've explored throughout this book, the strategic use of sound can enhance usability, evoke emotions, create product attachment, and lead to memorable experiences. Sound holds the power to provide a competitive advantage, helping products to stand out in a crowded marketplace. Integrating sound into inclusive design practices can also help ensure accessibility and usability for diverse audiences. By understanding the psychology of sound, employing rigorous testing methods, and asking the right questions, UX professionals can harness the power of sound to elevate their designs, mark out their product and brand as different from competitors, and produce memorable user experiences.

I've outlined here some key ways that sound should be considered separately from other aspects of UX—that is to say, that it has some special considerations that require additional attention from UX professionals. This knowledge is a work in progress, as researchers learn more about sound. When I started researching sound in the late 1990s, there was almost nothing written about sound design, beyond a small collection of film sound books. There were no school programs at universities that I was aware of that were undertaking research into sound design, and most general product design programs did not even think to include sound. Today, with the rise of sensory design practice, many designers realize the value that sound can bring to a product, and there is increasing awareness of the need for specialized training in sound. There are even dedicated degrees to sound design, and more researchers are beginning to explore our responses to sound.

The future of sonic user experience is full of potential, with emerging technologies offering new possibilities for immersive and personalized sound experiences. Looking ahead, we face both challenges and opportunities. Standardizing sound testing methods and navigating technological shifts will be crucial as we develop this nascent field. Consider sharing whatever methods and research questions that you end up using, to help us to expand this knowledge about sound as an important part of user experience.

Finally, I conclude with a list of some great sound user experience design to explore. I also encourage the reader to visit the website of the International Sound Awards to search award-winning examples (https://www.international-sound-awards.com/).

Apple iOS/Mac OS system sounds: Apple's system sounds have been a key part of its design language, designed to feel smooth, satisfying, and nonintrusive. From the clicking sound when an action is completed to the subtle notification sounds, Apple's system sounds are used not just for feedback, but to build a sense of identity and emotion. The user interface sounds are designed to be emotionally neutral but informative, with soft, rounded attack envelopes, minimal decay, and harmonic content designed to feel responsive but nonthreatening. The lock sound on iPhones is crisp and definitive, conveying security. Apple's sounds are clearly tuned to avoid harshness and instead aim for warmth, which complements its brand identity. For example, the ping of an incoming message or notification is designed to feel like a subtle, polite nudge rather than an alarming interruption. Sounds are designed with consistency across the operating system, contributing to a seamless experience where users feel the environment is familiar. The minimalist yet responsive sound design reinforces the overall tactile and visual experience of Apple products, making them feel polished, premium, and user-friendly.

Duolingo: Duolingo has created a highly engaging and gamified learning experience with a lot of focus on the audiovisual interaction. Their sound design is playful and encourages continued use with rewarding, character-driven sounds. Each time a user completes a lesson or a level, Duolingo plays a fun achievement sound. When users make mistakes or need to retry, they receive a distinctive sound, but it is designed not to be harsh or discouraging. Instead, it's gentle, offering a second chance. These sounds serve as rewards that help reinforce the user's progress and keep them motivated to continue learning. Duolingo's mascot, Duo the owl, frequently speaks through sound effects, such as quacking or laughing, depending on the user's actions. This gives the app a personal and light-hearted touch. The app's sound effects are designed to match its playful visual design, with upbeat, friendly tones that align with the brand's tone of voice. Explore how playful, character-driven sounds combining positive reinforcement with a friendly and approachable tone contribute to retention and motivation.

BMW electric vehicle soundscapes: BMW's electric vehicles do not have engine noise, so they partnered with Hans Zimmer to create a futuristic soundscape that represents the power and personality of their vehicles. The sounds are not merely for safety reasons; they add a layer of emotional and sensory engagement that would otherwise be absent, and enhance both the driving experience and the brand identity. They establish an emotional connection between the car and the driver while ensuring safety with clear, futuristic sonic cues. The sound design provides feedback that mimics the feeling of acceleration, engine power, and speed. These sounds evoke a sense of excitement and dynamism, creating an emotional connection to the driving experience. The soundscape also serves a functional role in alerting pedestrians to the vehicle's presence, offering a futuristic and calming sound rather than traditional car engine noise. BMW allows drivers to choose from different sound profiles, letting users select a driving experience that aligns with their personal tastes or mood. The car even emits sounds from different parts of the vehicle to create a three-dimensional effect.

Nintendo Switch: The Nintendo Switch is a great example of how sound can enhance the gaming experience without overwhelming it. The user interface sounds are fun, light, and closely connected to the hardware's tactile interactions. The Switch's user interface is filled with soft, playful sound effects that align with Nintendo's overall branding. Sounds are light-hearted, cute, and brief, designed to evoke a sense of joy and engagement in both children and adults. Whether users are docked or using the hand-held version, the sound design remains consistent, creating a unified user experience. The sound also works hand-in-hand with haptic feedback, so physical actions (like pressing a button or moving a joystick) are mirrored with satisfying sounds, making the device feel more responsive.

Mastercard: Mastercard has developed a unique audio brand identity with its iconic sound used in various interactions, including the app, commercials, and digital transactions, creating a unified experience that makes the brand recognizable. Mastercard worked with sound designers to create a signature sound logo, which is a short, memorable melody played when you make a payment, either in-app or via a card transaction. The sound is smooth and reassuring, designed to evoke a sense of security and reliability. The use of music helps Mastercard to humanize its brand and make the transaction experience feel less transactional and more engaging. It's designed to evoke positive feelings, reinforcing the sense of trust in the brand, which is crucial for financial services.

These examples highlight how sound can do so much more than just convey information: It can create and shape emotional responses, strengthen brand identities, guide and provide feedback for users, and enhance overall user experience. Each case study shows how thoughtful sound design can turn a basic interaction into something memorable, fun, or emotionally resonant.

It's my hope that this book begins a much-needed dialog with executives about sound in their products. Sound has so much to offer, and yet is so often overlooked, under-resourced and under-budgeted. To my readers, please share any user experience studies, methods, and knowledge if you are permitted, so that we may grow this field together and learn from each other.

I will conclude with a recommendation of a few podcasts, conferences, and organizations, which may be useful for someone interested in this field.

Podcasts

99% Invisible Podcast (https://99percentinvisible.org/): A design podcast that has many episodes about sound.

Google Design Podcasts (https://design.google/category/podcasts): A series of podcasts created by the Google Design team.

Twenty Thousand Hertz (https://www.20k.org/): Every episode of this excellent podcast is about an element of sound. I was interviewed on an episode about slot machine sound many years ago.

Conferences and Organizations

AudioMostly (https://audiomostly.com/): AudioMostly is the core annual sound conference that covers all aspects of sound as a design practice. While primarily academic, the conferences also often provide workshops and other hands-on activities. In 2025 the conference was co-hosted with the International Community for Auditory Display (ICAD).

Design Research Society (https://www.designresearchsociety.org/cpages/sigs): A broad organization dedicated to research aspects of design, with a sound-driven design special interest group.

International Sound Awards (https://www.international-sound-awards.com/): They call themselves the "Oscars of Sound," although it's open to everything from game sound, sound art, product sound, voice, and more, while primarily focused on audio branding. They maintain a directory of affiliated people, and are associated with the Audio Branding Society.

Acknowledgments

This book developed out of several years of consulting work for a few different companies, all of which came to me to ask, essentially, "What do we do with sound?" After nearly two decades of working on sound for film and games, as well as often swimming upstream in academia arguing for more inclusion of sound in digital media, design, games, and film programs, it should have been no surprise to me that sound was a bit of an afterthought in the corporate world. Sound was treated, I argued twenty years ago, as sort of an "ugly stepchild," and while things are improving, there is still work to be done. This book is not meant to be the final word on sound in UX, but rather the beginnings of a larger conversation and much-needed discussion. I hope it challenges assumptions, provokes questions, sparks debate, and encourages deeper exploration and further research into how sound shapes user experience in ways we have yet to fully understand.

The book could not have been written without the often-long discussions I have had with many sound friends over the last few years. Special thanks to Peter "Pdx" Drescher, now retired, who got furious at my treating sound as if its impacts could be tested and quantified in this manner and forced me to carefully reevaluate many of my assumptions. Thanks to Andrew Scott Duncan—sound designer and composer with whom I worked for some time over the last few years and during the writing of this book; our conversations about sound often spilled over from meetings into wider discussions about sound's purpose—whose input on earlier drafts was invaluable. Thanks to good friends and sounding boards Damian Kastbauer (software product manager at Audiokinetic) and Scott Arnold (director of design systems, design ops and UX strategy at Motorola), and the teams at the companies I've consulted with who have hired me to help them understand sound. Thanks also to the detailed feedback of the anonymous reviewers, and to my editors past and present at the MIT Press, Doug Sery and now Noah Springer.

Thanks to my colleagues and students at the School of Information Technology at Carleton University in Ottawa, Canada, including especially those who worked with me on research that often developed from questions that came up as I was writing this book, especially Hannah Johnston, Adel Manji, Maya Murad, and Miguel Valle. My gratitude always goes out to my collaborators and academic colleagues past and present, particularly Bill Kapralos and Paul Théberge. Part of the research work that provided the backbone of some of the ideas presented was funded by the Social Sciences and Humanities Research Council of Canada in the form of an Insight Grant and an Insight Development Grant. Finally, my gratitude to Philip Tagg, for teaching me to tune into the noise others tuned out. He would have hated my bringing these ideas to the corporate world, but I hope he would have forgiven me for trying to make our world sound a little better.

Glossary

Introduction

Aesthetic-usability effect is a phenomenon where users perceive aesthetically pleasing designs as easier to use—even if they aren't more functional—highlighting the strong influence of visual appeal on user experience.

Anthropomorphism is the attribution of human traits, emotions, or intentions to nonhuman entities, such as robots, interfaces, or sounds, often used to make technology feel more relatable or engaging.

Easter eggs are secret features, hidden messages, or inside jokes embedded in products. A common sound Easter egg found in film is the "Wilhelm scream," for example.

Ocularcentrism is the tendency in a culture to prioritize vision over other senses in how we perceive, understand, and design the world, often leading to visual-dominant interfaces or research that neglects sound, touch, or other sensory inputs.

Skeuomorphism is a design approach of digital interfaces in which objects mimic their real-world counterparts, such as a digital desktop sticky note looking and working like a real-world sticky note. Sonic examples are the click-wind of taking a photograph or the keypad clicks on a phone.

Sound quality evaluation a traditional area of product design from an engineering perspective that evaluates the fidelity of the sound being produced, such as if a tweak in the shape of an electric guitar may improve the sound perception. The field often also seeks to reduce extraneous noise from a product.

Chapter 1

Connotations refer to the emotional, cultural, or associative meanings a sound evokes beyond its literal function. For example, a ticking clock might connote urgency or anxiety, depending on context.

Multimodality is the use and integration of multiple sensory or communicative modes—such as visual, auditory, tactile, and textual—in interaction or design. Multimodality recognizes that people interpret meaning not from a single channel (or mode), but through the combined effect of multiple inputs.

Sensory memory is short-term spontaneous memory that holds sensory information—such as sights, sounds, or smells—for a few seconds or less. It acts as a filter before information moves to short-term memory, and plays a role in how users notice, react to, or become conditioned by sensory stimuli.

Chapter 2

Auditory icons are sounds that are designed to represent objects, functions, or events. Auditory icons exhibit a clear and intentional resemblance to the real-world sounds or concepts they represent, such as the use of the camera click sound on mobile phones. They are therefore usually *skeuomorphic*.

Autonomous sensory meridian response (ASMR) is a tingling, calming sensation often felt on the scalp, neck, or spine in response to specific auditory or visual triggers—like whispering, tapping, or soft crinkling sounds. ASMR content is widely used for relaxation and stress relief, has become a niche but influential area in sensory media, and is often evoked through close-miking sounds.

Earcons consist of sequences of musical notes or patterns used to convey specific information or actions. Earcons are typically composed to be distinctive and easily recognizable, allowing users to develop a learned association of the sound with specific meanings or action.

Mnemonics are sound cues or musical elements used to aid memory and brand recognition. In UX and marketing, this often refers to a *sonic logo*—a short, distinctive sound (such as Intel's chime or Netflix's "ta-dum") designed to create a strong mental association with a product or brand.

Muzak, originally a brand name, is now a general term for unobtrusive background music used in public or commercial spaces (such as elevators, malls, or waiting rooms). Muzak is often engineered to influence mood or behavior subtly, such as encouraging shoppers to linger longer or feel more relaxed.

Sensory design is the intentional use of sensory elements—sound, touch, sight, smell, and taste—in product or environmental design to shape user experience, emotion, and behavior. It emphasizes designing *with* the senses, not just for function or aesthetics.

Servicescapes are the physical or sensory environments in which a service experience takes place—including layout, sound, lighting, textures, and smells. A well-designed servicescape can influence customer satisfaction, emotional response, and even perceived service quality.

Social surrogacy is a term for how individuals use objects or media as a substitute for social interaction, companionship, or emotional connection when direct human interaction is limited or unavailable. Media or sensory experiences (e.g., music, familiar voices, or even the sound of a TV) can fulfill social needs and reduce feelings of loneliness. Sound-based interfaces or content can act as a form of emotional companionship or comfort.

Transscansion a term created by musicologist Philip Tagg to refer to a series of musical notes or sound effects that stand in for the prosody of speech of syllables in a word or phrase in such a way that the repetition of these sounds may offer a substitution for the actual word(s) in our minds, and often created by repeating the syllables in text on a screen.

Unboxing is the experience of opening and engaging with a product for the first time, often designed to be multisensory and emotionally rewarding. Unboxing is a critical moment of brand storytelling—through packaging texture, sound (like the snap of a seal or rustle of paper), and visual presentation.

Chapter 3

Acoustic startle reflex is an involuntary, rapid physical reaction (like blinking or muscle tensing) triggered by a sudden loud sound. It's a protective response of the nervous system and can be useful in measuring arousal, fear, or attention in experiments.

Auditory adaptation is a reduction in sensitivity to a constant or repetitive sound over time. For example, background noise like a fan or hum may initially be noticeable but fades from conscious awareness—a key concept in sound design for environments that will be experienced over the long term.

Cocktail party effect refers to the brain's ability to focus on a specific auditory source (e.g., one voice) in a noisy environment, such as a party or crowded room (hence the name "cocktail party"). It demonstrates selective attention in auditory perception and has implications for designing voice interfaces or alert systems.

Colavita effect is a perceptual bias where people tend to prioritize visual stimuli over auditory ones when both are presented simultaneously. This effect is important in multimodal interface design, highlighting the challenge of ensuring that sound is noticed or integrated effectively.

Earworms are catchy fragments of music or jingles that involuntarily repeat in a person's mind. Often used intentionally in branding (e.g., sonic logos), earworms reflect how music can persist in memory and influence mood or behavior.

Mere exposure effect is a psychological phenomenon where repeated exposure to a stimulus increases a person's preference for it. In sound design, this can explain why users grow to like certain interface tones, notification chimes, or musical themes over time.

Multimedia effect is the enhancement of learning or memory when information is presented using multiple sensory modalities, such as combining visuals with sound or text. Effective multimedia design aligns different modes to reinforce understanding rather than overwhelm. For example, we can provide task-relevant information through sound effects to help someone learn and navigate a new piece of software.

Pseudo-haptics refers to the illusion of touch or tactile feedback created through non-tactile means, such as sound or visual cues. For example, audio cues synchronized with visual motion can simulate resistance or texture in virtual environments without actual haptic devices.

Sensation transference is when the qualities of a product's sensory features (e.g., its sound, feel, or taste) are unconsciously transferred to the perception of the product itself. For example, a heavy, solid-sounding click might make a digital product feel more premium or trustworthy.

Chapter 4

Foley are custom-recorded sound effects that sync with on-screen actions to enhance realism, such as footsteps, cloth movement, or object handling. Named after Jack Foley, this technique is a cornerstone of film and game audio production. Sometimes people use the word "Foley" now to refer to any custom-recorded sounds, as opposed to sounds from a stock sound library.

Low-fi, short for low fidelity, refers to sound that deliberately retains or adds imperfections, noise, or degradation. In sound design, low-fi can convey intimacy, nostalgia, or a handcrafted feel.

LUFS (loudness units relative to full scale) is a standardized unit for measuring perceived loudness over time by averaging peak levels, used in audio mastering to ensure consistent playback levels across platforms. LUFS is more aligned with human hearing than traditional peak or root mean square (RMS) levels.

Monophonic sound is a sound recording or playback format where all audio signals are mixed into a single channel, as opposed to stereo or surround sound. Mono sound has no spatial dimension but is still used for simplicity or stylistic effect.

Stems are individual sub-mixes of grouped audio elements (e.g., vocals, drums, synthesizers) exported separately to allow for flexible remixing, mastering, or integration with other media.

Temp tracks are temporary audio or music files placed in a film, animation, or media project during early edits to guide mood or pacing. While useful for communication, temp tracks can cause *temp love*, where creators become overly attached and resist original compositions.

Chapter 6

Cognitive walkthroughs are a usability evaluation method where experts step through a task as if they were the user, asking questions at each step about ease of learning and likelihood of user success. It focuses on identifying design issues that affect first-time users.

Focus groups are a moderated discussion with a small group of target users to gather opinions, feelings, and perceptions about a product, idea, or experience. Focus groups are useful in early design phases or to interpret user reactions qualitatively.

Personas are fictional characters created to represent a specific user type based on research data. Personas help designers empathize with users' goals, behaviors, and frustrations, and serve as useful reminders or touchpoints of who they are designing for. *User profiles*, on the other hand, are summaries of a *real* user's characteristics, preferences, and usage context. Unlike personas, user profiles are often drawn from actual data and can be updated dynamically, especially in interactive or adaptive systems.

Think-aloud is a research method where participants verbalize their thoughts while interacting with a product or interface. This provides insight into cognitive processes, expectations, and frustrations that aren't always visible through behavior alone

Chapter 7

Arousal in the scientific literature refers to a state of physiological and psychological activation ranging from drowsiness to alertness. In UX and sound research, arousal is often measured to understand emotional intensity or engagement, such as how a sudden sound might spike attention.

Chapter 8

Affect misattribution procedure is a paradigm that illustrates the misattribution of affective information based on the contexts in which they were experienced.

Attention checks are simple questions or tasks inserted into a study or survey to ensure participants are paying attention (e.g., "Select option 3 for this question"). It helps filter out careless or noncompliant responses, especially in online research.

Confirmation bias is the tendency to search for, interpret, and remember information that confirms existing beliefs, while ignoring contradicting evidence. It affects user feedback interpretation, design evaluation, and even research framing.

Ear fatigue, or *listening fatigue,* is the mental or sensory exhaustion that results from prolonged or effortful listening—especially in noisy environments or with poor audio quality. It's relevant in UX for interface design, assistive tech, and any long-form audio experiences.

Groove is the sense of rhythmic feel or musical entrainment that makes you want to move or tap along with music. Groove is not just a musical term; it's tied to motor resonance and pleasure, and is increasingly studied in cognitive science and human–computer interaction.

Psychographics refers to qualitative user data that capture lifestyle, attitudes, values, interests, and personality traits. Unlike demographics, psychographics help build deeper user insights and are often used in persona creation or targeted design.

Social desirability bias is the tendency for participants to respond in a way they think is socially acceptable or favorable, rather than truthfully. This can skew results in self-reported data, especially around topics such as behavior, preferences, or identity.

Sunk cost fallacy refers to a cognitive bias where people continue an investment (of time, money, or effort) due to what they've already put in, even when it no longer makes sense. This is relevant in UX when users resist abandoning poor experiences because they've "already come this far."

Chapter 9

Onboarding is the process of introducing users to a product or system and helping them understand how to use it effectively. Good onboarding can involve tutorials, tool tips, sound cues, or feedback mechanisms to ease learning and reduce drop-off.

Notes

Introduction

1. "No More Waking the Wife," Reddit post from r/playstation by Ikewiththebeard, 2023. Grammar and punctuation edited for readability: https://www.reddit.com/r/playstation/comments/15eb7ki/no_more_waking_the_wife/.

Chapter 2

1. CNBC video on TikTok: https://www.tiktok.com/@cnbcmakeit/video/7140749522816519470?lang=en.

2. From the *Method Podcast* from Google Design, 2018: https://design.google/library/method-podcast-episode-1.

3. E.g., AppleCare/Apple Support Hold Music playlist by "tony": https://open.spotify.com/playlist/4RWicgj2hk2cGVrdfIQZl8.

4. The One Co Cheshire, Facebook post, December 5, 2022, https://www.facebook.com/watch/?v=843281973591643.

Chapter 3

1. It should be noted that silence can also work to capture attention, particularly when used in conjunction with periods of background music, where a sudden drop to silence in an advertisement, for instance, can increase attention and focus.

Chapter 4

1. BMW even created their own "museum of sounds" of their cars, recording every sound of the car for an NFT collection: https://www.branding.news/2022/02/14/nft-collection-captures-bmw-m-engines-sounds-so-that-brands-fans-can-enjoy-them/.

2. From the *Method Podcast* from Google Design, 2018, https://design.google/library /method-podcast-episode-1.

3. "No More Waking the Wife," Reddit post on r/playstation by Ikewiththebeard, 2023, https://www.reddit.com/r/playstation/comments/15eb7ki/no_more_waking_the_wife/.

4. Peter Drescher, personal correspondence, May 2023.

Chapter 5

1. Note: There are usually regulations. In the United States, they must sound for at least eight seconds. See the National Highway Traffic Safety Administration interpretation ID: 22058: https://www.nhtsa.gov/interpretations/22058.

References

Abraham, Alon, and Vivian E. Drory. 2014. "Listening to music during electromyography does not influence the examinee's anxiety and pain levels." *Muscle & Nerve* 50(3): 445–447. https://doi.org/10.1002/mus.24291.

Adelstein, B. D., D. R. Begault, M. R. Anderson, and E. M. Wenzel. 2003. "Sensitivity to haptic-audio asynchrony." *Proceedings of the International Conference on Multimodal Interfaces*, 73–76. Vancouver, Canada, November 5–7. https://doi.org/10.1145/958432.958448.

Adkins, Frankie. 2021. "The real reason hold music bothers you so much." *Wired*, December 6. https://www.wired.co.uk/article/hold-music-behavior.

Albert, William, and E. Dixon. 2003. "Is this what you expected? The use of expectation measures in usability testing." *Proceedings of the Usability Professionals Association 2003 Conference*, Scottsdale, AZ, June 25.

Alexomanolaki, Margarita, Chris Kennett, and Catherine Loveday. 2010. "Music as first-order and second-order conditioning in TV commercials." *Music and the Moving Image* 3(2): 39–50.

Almiron, Paula, Francisco Barbosa Escobar, Abhishek Pathak, Charles Spence, and Carlos Velasco. 2021. "Searching for the sound of premium beer." *Food Quality and Preference* 88: 104088.

Altinsoy, M. Ercan, and Ute Jekosch. 2012. "The semantic space of vehicle sounds: Developing a semantic differential with regard to customer perception." *Journal of the Audio Engineering Society* 60(1–2): 13–20.

Anglada-Torta, Manuel, Kerry Schofield, Tabitha Trahan, and Daniel Müllensiefen. 2022. "I've heard that brand before: The role of music recognition on consumer choice." *International Journal of Advertising* 41(8): 1567–1587.

Anzenbacher, Christoph. 2013. "Audiologos: Are there common rules for the design of a catchy melodic figure?" In *Audio Branding Academy Yearbook 2011–2012*, edited by K. Bronner, R. Hirt, and C. Ringe, 129–140. Baden-Baden: Nomos Verlagsgesellschaft mbH.

Ara, Alberto, and Josep Marco-Pallarés. 2021. "Different theta connectivity patterns underlie pleasantness evoked by familiar and unfamiliar music." *Scientific Reports* 11: 18523.

Asutay, Erkin, Daniel Västfjäll, Ana Tajadura-Jiménez, Anders Genell, Penny Bergman, and Mendel Kleiner. 2012. "Emoacoustics: A study of the psychoacoustical and psychological dimensions of emotional sound design." *Journal of the Audio Engineering Society* 60(1–2): 21–28.

Azpiri, Jon. 2022. "A musical experiment from the '80s to keep B.C. teens from loitering is now a global practice." *CBC News*, November 12. https://www.cbc.ca/news/canada/british-columbia/muzak-7-eleven-stores-bc-80s-1.6646742.

Ballas, James A. 2007. "Self-produced sound: Tightly binding haptics and audio." *Proceedings of the International Workshop on Haptic and Audio Interaction Design*. Seoul, Korea, November 29–30.

Barrass, Stephen. 1996. "EarBenders: Using stories about listening to design auditory interfaces." *Proceedings of the First Asia-Pacific Conference on Human Computer Interaction APCHI*, Rotorua, New Zealand, June 29–July 2.

Barreto, Armando, Jing Zhai, and Malek Adjouadi. 2007. "Non-intrusive physiological monitoring for automated stress detection in human-computer interaction." *International Workshop on Human-Computer Interaction HCI 2007*, Rio de Janeiro, Brazil, October 20.

Barrett, Lisa Feldman, Batja Mesquita, Kevin N. Ochsner, and James J. Gross. 2007. "The experience of emotion." *Annual Review of Psychology* 58: 373–403.

Bech, Søren. 1992. "Selection and training of subjects for listening tests on sound reproducing equipment." *Journal of the Audio Engineering Society* 40: 590–610.

Bech, Søren, and Nick Zacharov. 2006. *Perceptual Audio Evaluation: Theory, Method and Application*. West Sussex: John Wiley & Sons.

Beckerman, Joel. 2014. *The Sonic Boom: How Sound Transforms the Way We Think, Feel and Buy*. Boston: Mariner Books.

Bell, Lynne, Julia Vogt, Cesco Willemse, Tim Routledge, Laurie T. Butler, and Michiko Sakaki. 2018. "Beyond self-report: A review of physiological and neuroscientific methods to investigate consumer behavior." *Frontiers in Psychology* 9: 1655. https://doi.org/10.3389/fpsyg.2018.01655.

Belova, Mariia, Dmitry Dimitriev, Aleksei Dimitriev, and Olga Indeykina. 2019. "The effect of sounds on nonlinear measures of heart rate variability." *FASEB Journal* 33: 533–510. https://doi.org/10.1096/fasebj.2019.33.1_supplement.533.10.

Benghanem, Abdelghani, Olivier Valentin, Philippe-Aubert Gauthier, and Alain Berry. 2021. "Sound quality of side-by-side vehicles: Investigation of multidimensional sensory profiles and loudness equalization in an industrial context." *Acta Acustica* 5(7). https://doi.org/10.1051/aacus/2020032.

Berger, Christopher C., and H. Henrik Ehrsson. 2013. "Mental imagery changes multisensory perception." *Current Biology* 23: 1367–1372. https://doi.org/10.1016/j.cub.2013.05.048.

Bergman, Penny, Daniel Västfjäll, Ana Tajadura-Jiménez, and Erkin Asutay. 2016. "Auditory-induced emotion mediates perceptual categorization of everyday sounds." *Frontiers in Psychology* 7: 1565. https://doi.org/10.3389/fpsyg.2016.01565.

Bettiga, Debora, Lucio Lamberti, and Giuliano Noci. 2017. "Do mind and body agree? Unconscious versus conscious arousal in product attitude formation." *Journal of Business Research* 75: 108–117.

Bhatara, Anjali, Xavier Dartis, David Morizet, and Sébastien Romagny. 2025. "Designing emotional and intuitive sounds for tech: Insights from psychoacoustics." *Human Behavior and Emerging Technologies* 2025: 5925146. https://doi.org/10.1155/hbe2/5925146.

Bi, Youyi, Tahira Reid, and Patricia Davies. 2017. "An exploratory study on proposed new sounds for future products." *Noise Control Engineering Journal* 65(3): 244–260.

Biocca, Frank, Yasuhiro Inoue, Andy Lee, Heather Polinsky, and Arthur Tang. 2002. "Visual cues and virtual touch: Role of visual stimuli and intercessory integration in cross-modal haptic illusions and the sense of presence." *Proceedings of Presence*, Porto, Portugal, October 9–11.

Biswas, Dipayan, Kaisa Lund, and Courtney Szocs. 2018. "Sounds like a healthy retail atmospheric strategy: Effects of ambient music and background noise on food sales." *Journal of the Academy of Marketing Science* 47: 37–55. https://doi.org/10.1007/s11747-018-0583-8.

Blain, Salomé, Francesca Talamini, Lesly Fornoni, Aurélie Bidet-Caulet, and Anne Caclin. 2022. "Shared cognitive resources between memory and attention during sound-sequence encoding." *Attention, Perception, & Psychophysics* 84(3): 739–759.

Bolinger, Dwight. 1978. "Intonation across languages." *Universals of Human Language* 2: 471–524.

Bolls, Paul D. 2006. "It's just your imagination: The effect of imagery on recognition of product- versus non-product-related information in radio advertisements." *Journal of Radio Studies* 13(2): 201–213.

Bolls, Paul D., and Annie Lang. 2003. "I saw it on the radio: The allocation of attention to high-imagery radio advertisements." *Media Psychology* 5(1): 33–55.

Bonneel, Nicolas, Clara Suied, Isabelle Viaud-Delmon, and George Drettakis. 2010. "Bimodal perception of audio-visual material properties for virtual environments." *ACM Transactions on Applied Perception* 7(1): 1–16.

Boone, R. Thomas, and Joseph G. Cunningham. 2001. "Children's expression of emotional meaning in music through expressive body movement." *Journal of Nonverbal Behavior* 25: 21–41.

Borowiak, Adam, Ulrich Reiter, and U. Peter Svensson. 2014. "Momentary quality of experience: Users' audio quality preferences measured under different presentation conditions." *Journal of the Audio Engineering Society* 62(4): 235–243.

Bowles, Sarah Parker. 2013. "Stitch perfect." *Harper's Bazaar*, November, 97–98.

Bradley, Margaret M., and Peter J. Lang. 2000. "Affective reactions to acoustic stimuli." *Psychophysiology* 37(2): 204–215.

Brattico, Elvira, Karen Johanne Pallesen, Olga Varyagina, Christopher Bailey, Irina Anourova, Miika Järvenpää, Tuomas Eerola, and Mari Tervaniemi. 2009. "Neural discrimination of nonprototypical chords in music experts and laymen: An MEG study." *Journal of Cognitive Neuroscience* 21(11): 2230–2244.

Bresin, Roberto, and Anders Friberg. 2001. "Expressive musical icons." *Proceedings of the 2001 International Conference on Auditory Display*, Espoo, Finland, July 29–August 1.

Bresin, Roberto, Anna de Witt, Stefano Papetti, Marco Civolani, and Federico Fontana. 2010. "Expressive sonification of footstep sounds." *Proceedings of iSon 2010*, Stockholm, Sweden, April 7.

Brewster, Stephen. *Providing a Model for the Use of Sound in User Interfaces*. University of York, Department of Computer Science, 1992.

Brewster, Stephen A., Peter C. Wright, and Alistair D. N. Edwards. 1993. "An evaluation of earcons for use in auditory human-computer interfaces." *Proceedings of the INTERACT'93 and CHI'93 conference on Human Factors in Computing Systems*, Amsterdam, April 24–29, 222–227.

Brown, Phil, Chris Forreseter, Peter Otto, David Roach, Anastasia Devana, Ty Kingsmore, Jack Joseph Puig, and Will Wolcott. 2016. "Always be closing (This isn't marketing after all)." *Proceedings of the Twenty-First Annual Interactive Audio Conference Project BAR-B-Q 2016*. Berne, TX, October 12–14. https://www.projectbarbq.com/reports/.

Bruun, Anders, and Simon Ahm. 2015. "Mind the gap! Comparing retrospective and concurrent ratings of emotion in user experience evaluation." *Proceedings of the 15th Human-Computer Interaction (INTERACT) Conference*, Bamberg, Germany, September 14–18, 237–254.

Buchwald, Mikołaj, Szymo Kupiński, Adam Bykowski, Joanna Marcinkowska, Dawid Ratajczyk, and Marcin Jukiewicz. 2019. "Electrodermal activity as a measure of cognitive load: A methodological approach." *IEEE Signal Processing: Algorithms, Architectures, Arrangements, and Applications (SPA)*, Poznan, Poland, September 18–20, 175–179.

Bull, James. 2009. "The narrative ear: Realities in creating sound, story and emotion for brands." In *Audio Branding: Brands, Sound and Communication*, edited by Kai Bronner and Rainer Hirt, 205–214. Baden-Baden: Nomos Verlagsgesellschaft mbH.

Byron, Ellen. 2012. "The search for sweet sounds that sell: Household products' clicks and hums are no accident; Light piano music when the dishwasher is done?" *Wall Street Journal*, October 23. https://www.wsj.com/articles/SB1000142405297020 3406404578074671598804116.

Caetano, Gina. 2007. "Brain mechanisms of audiotactile and audiomotor interactions." PhD thesis, Helsinki University of Technology.

Cafaro, Angelo, Hannes Högni Vilhjálmsson, and Timothy Bickmore. 2016. "First impressions in human–agent virtual encounters." *ACM Transactions on Computer-Human Interaction* 23(4): 1–40. https://doi.org/10.1145/2940325.

Cañadas-Quesada, F. J., and A. Reyes-Lecuona. 2006. "Improvement of perceived stiffness using auditory stimuli in haptic virtual reality." *MELECON 2006: 2006 IEEE Mediterranean Electrotechnical Conference*, Málaga, Spain, May 16–19, 462–465.

Caruelle, Delphine, Anders Gustafsson, Poja Shams, and Line Lervik-Olsen. 2019. "The use of electrodermal activity (EDA) measurement to understand consumer emotions: A literature review and a call for action." *Journal of Business Research* 104: 146–160.

Carvalho, Felipe Reinoso, Qian Janice Wang, Raymond Van Ee, Dominique Persoone, and Charles Spence. 2021. "'Smooth operator': Music modulates the perceived creaminess, sweetness, and bitterness of chocolate." *Appetite* 108: 383–390.

Caviola, Sara, Chiara Visentin, Erika Borella, Irene Mammarella, and Nicola Prodi. 2021. "Out of the noise: Effects of sound environment on maths performance in middle-school students." *Journal of Environmental Psychology* 73: 101552.

Cespedes-Guevara, Julian, and Nicola Dibben. 2022. "The role of embodied simulation and visual imagery in emotional contagion with music." *Music & Science* 5: 1–27.

Cespedes-Guevara, Julian, and Tuomas Eerola. 2018. "Music communicates affects, not basic emotions: A constructionist account of attribution of emotional meanings to music." *Frontiers in Psychology* 9: 215. https://doi.org/10.3389/fpsyg.2018.00215.

Chandrasekera, Tilanka, So-Yeon Yoon, and Newton D'Souza. 2015. "Virtual environments with soundscapes: A study on immersion and effects of spatial abilities." *Environment and Planning B: Planning and Design* 42: 1003–1019.

Chen, Fangyuan, Jaideep Sengupta, and Rashmi Adaval. 2018. "Does endowing a product with life make one feel more alive? The effect of product anthropomorphism on consumer vitality." *Journal of the Association for Consumer Research* 3(4): 503–513.

Choe, Chong S., Robert B. Welch, Robb M. Gilford, and James F. Juola. 1975. "The 'ventriloquist effect': Visual dominance or response bias?" *Perception & Psychophysics* 18(1): 55–60.

Chu, Chih-Hsing, and Erh-Ting Kao. 2020. "A comparative study of design evaluation with virtual prototypes versus a physical product." *Applied Sciences* 10(14): 4723. https://doi.org/10.3390/app10144723.

Cian, Edoardo, Daniele Marini, and Anne Maass. 2022. "The sound of social class: Do music preferences signal status?" *Psychology of Music* 50(3): 960–975.

Clausen, Noel, Jasmin Stein, Katharina Zwosta, and Anja Kräplin. 2023. "Examining the relationship between task-based self-control performance and different cortisol response profiles." *OSF*, May 10. https://doi.org/10.17605/OSF.IO/B5G73.

Colavita, Francis B. 1974. "Human sensory dominance." *Perceptual Psychophysics* 16(2): 409–412.

Collins, Karen. 2003. "The future is happening already: Industrial music, dystopia and the aesthetic of the machine." PhD thesis, University of Liverpool.

Collins, Karen. 2015. "Sonic subjectivity and auditory perspective in *Ratatouille*." *Animation Journal* 8(3): 283–299.

Collins, Karen. 2016. "Game sound in the mechanical arcades: An audio archaeology." *Game Studies* 16(1). http://gamestudies.org/1601/articles/collins.

Collins, KC, and Hannah Johnston. 2023. "A free verbalization method of evaluating sound design: The effectiveness of artificially intelligent natural language processing methods and tools." *ACM AudioMostly*. Edinburgh, August 30–September 1.

Collins, KC, and Adel Manji. 2024. "Humanizing AI-generated music: Do listeners hear a difference?" *Audio Engineering Society Convention 156*, Madrid, Spain, June 14–17.

Collins, KC, Adel Manji, Hannah Johnston, and Bill Kapralos. 2024. "Presentation mode and its impact on sentiment in free verbalization responses to sounds." *Audio Engineering Society Convention 156*, Madrid, Spain, June 14–17.

Collins, KC, Maya Murad, and Adel Manji. 2025. "Sound's influences on perceptions of anthropomorphism and animacy in interfaces." *ACM AudioMostly'25*, Coimbra, Portugal.

Cooley, Millicent. 1998. "Sound+image in design." *Proceedings of ICAD '98*. Glasgow, November 1–4.

Cox, Arnie. 2001. "The mimetic hypothesis and embodied musical meaning." *Music Science* 5(2): 195–212.

Craig, David, and Stuart Cunningham. 2017. "Toy unboxing: Living in a(n unregulated) material world." *Media International Australia* 163(1): 78–86. https://doi.org/10.1177/1329878X17693700.

Cuadra, Andrea, Jessica Bethune, Rony Krell, Alexa Lempel, Katrin Hänsel, Armin Shahrokni, Deborah Estrin, and Nicola Dell. 2023. "Designing voice-first ambient interfaces to support aging in place." *Proceedings of the 2023 ACM Designing Interactive Systems Conference*, Pittsburgh, PA, July 10–14, 2189–2205.

Cunningham, Stuart, Iain McGregor, Jonathan Weinel, John Darby, and Tony Stockman. 2023. "Towards a framework of aesthetics in sonic interaction." *Audio Mostly 2023*, Edinburgh, August 30–September 1. https://doi.org/10.1145/3616195.3616219.

Dahl, Sofia. 2005. "On the beat: Human movement and timing in the production and perception of music." PhD thesis, KTH Stockholm.

Date, Yuki, Shunsuke Ishimitsu, and Natsuki Yamagiwa. 2019. "Improvement of heart rate variability analysis for sound design." *ICIC Express Letters Part B: Applications* 10(2): 159–166.

Davidson, Austin, and Justin Robert Keene. 2019. "Alone in a crowded room: The impact of crowd noise in music recordings on social facilitation, mental imagery, and cognitive processing." *Journal of Radio & Audio Media* 26(2): 299–317.

Davis, Michael, David L. Walker, and Younglim Lee. 1997. "Roles of the amygdala and bed nucleus of the stria terminalis in fear and anxiety measured with the acoustic startle reflex." *Annals of the New York Academy of Sciences* 821: 305–331.

Day, Robina A., William Forde Thompson, and Simon Boag. 2020. "Characterizing experiences of music-evoked visual imagery in high prevalence contexts." *Psychomusicology: Music, Mind, and Brain* 30(2): 72–87.

De Borst, Aline W., Giancarlo Valente, Iiro P. Jääskeläinen, and Pia Tikka. 2016. "Brain-based decoding of mentally imagined film clips and sounds reveals experience-based information patterns in film professionals." *NeuroImage* 129: 428–438.

Delle Monache, Stefano, and Davide Rocchesso. 2019. "Sketching sonic interactions." *Foundations in Sound Design for Embedded Media: A Multidisciplinary Approach* (2019): 79–101.

Derks-Dijkman, Marije W., Rebecca S. Schaefer, and Roy P. C. Kessels. 2023. "Musical mnemonics in cognitively unimpaired individuals and individuals with Alzheimer's dementia: A systematic review." *Neuropsychology Review* 34(2): 455–477. https://doi.org/10.1007/s11065-023-09515-6.

Díaz, Iñaki, Josune Hernantes, Ignacio Mansa, Alberto Lozano, Diego Borro, Jorge Juan Gil, and Emilio Sánchez. 2006. "Influence of multisensory feedback on haptic accessibility tasks." *Virtual Reality* 10: 31–40.

Dickson, Crispin, and Karl Bolin. 2014. "Continuous judgment by category-ratio scaling of aircraft noise." *Applied Acoustics* 84: 3–8.

DiFranco, David E., G. Lee Beauregard, and Mandayam A. Srinivasan. 1997. "The effect of auditory cues on the haptic perception of stiffness in virtual environments." *ASME International Mechanical Engineering Congress and Exposition* 18244: 17–22.

Diroll, Anne, and Diana Hlebovy. 2003. "Inverse relationship between blood volume and blood pressure." *Nephrology Nursing Journal* 30(4): 460–461.

Dixon, Mike J., Karen Collins, Kevin A. Harrigan, Candice Graydon, and Jonathan A. Fugelsang. 2013. "Using sound to unmask losses disguised as wins in multiline slot machines." *Journal of Gambling Studies* 31(1): 1–15.

Dixon, Mike J., Kevin A. Harrigan, Rajwant Sandhu, Karen Collins, and Jonathan A. Fugelsang. 2010. "Losses disguised as wins in modern multi-line video slot machines." *Addiction* 105(10): 1819–1824. https://doi.org/10.1111/j.1360-0443.2010.03050.

Dixon, Mike J., Jeff Templeton, Karen Collins, Lisa Wojtowicz, Kevin A. Harrigan, Jonathan A. Fugelsang, and Vivian Siu. 2015. "Exploring attention in the 'reel' world: Visual and auditory influences on near-misses in multi-line slot machine play." In *The Handbook of Attention,* edited by J. Fawcett, E. F. Risko, and A. Kingstone, 549–574. Cambridge, MA: MIT Press.

Doucé, Lieve, Carmen Adams, Olivia Petit, and Anton Nijholt. 2022. "Crossmodal congruency between background music and the online store environment: The moderating role of shopping goals." *Frontiers in Psychology* 13: 883920.

Dousty, Mehdy, Sabalan Daneshvar, and Majid Haghjoo. 2011. "The effects of sedative music, arousal music, and silence on electrocardiography signals." *Journal of Electrocardiology* 44(3): 396e1–e6.

Doyle, Peter. 2004. "From 'My Blue Heaven' to 'Race with the Devil': Echo, reverb and (dis)ordered space in early popular music recording." *Popular Music* 23(1): 31–49.

Drescher, Peter. 2008. "Could ringtones be more annoying?" In *From Pac-Man to Pop Music,* edited by K. Collins. Aldershot: Ashgate.

Drescher, Peter. 2011. "Generating audio UI." *Mobile User Experience (MEX)* conference presentation, London.

Duggal, Simon. 2024. "Metering: Peak, RMS and LUFS." In *Record, Mix and Master: A Beginner's Guide to Audio Production,* edited by Simon Duggal and Paul J. Rogers, 243–246. Cham: Springer Nature Switzerland.

Eerola, Tuomas, James Armitage, Nadine Lavan, and Sarah Knight. 2021. "Online data collection in auditory perception and cognition research: Recruitment, testing, data quality, and ethical considerations." *Auditory Perception & Cognition* 4(3–4): 251–280. https://doi.org/10.1080/25742442.2021.2007718.

Egermann, Hauke, Mary Elizabeth Sutherland, Oliver Grewe, Frederik Nagel, Reinhard Kopiez, and Eckart Altenmüller. 2011. "Does music listening in a social context alter experience? A physiological and psychological perspective on emotion." *Musicae Scientiae* 15(3): 307–323. https://doi.org/10.1177/1029864911399497.

Epley, Nicholas, Adam Waytz, and John T. Cacioppo. 2007. "On seeing human: A three-factor theory of anthropomorphism." *Psychological Review* 114(4): 864–886.

Fairclough, Stephen H., and Louise Venables. "Prediction of subjective states from psychophysiology: A multivariate approach." *Biological Psychology* 71, no. 1 (2006): 100-110.

Falkenberg, Kjetil, Emma Frid, Martin Ljungdahl Eriksson, Tobias Otterbring, and Sven-Olov Daunfeldt. 2021. "Auditory notification of customer actions in a virtual retail environment: Sound design, awareness and attention." *International Conference on Auditory Display*, virtual conference, June 25–28.

Fastl, Hugo. 2005. "Psycho-acoustics and sound quality." In *Communication Acoustics*, edited by Jens Blauert. Berlin: Springer. https://doi.org/10.1007/3-540-27437-5_6.

Fela, Randy Frans, Nick Zacharov, and Søren Forchhammer. 2022. "Assessor selection process for perceptual quality evaluation of 360 audiovisual content." *Journal of Audio Engineering Society* 70(10): 824–842. https://doi.org/10.17743/jaes.2022.0037.

Fernström, Mikael, Liam J. Bannon, and Eoin Brazil. 2004. "An investigation of soft-button widgets using sound." *Le journées de design sonore*. Paris, October 13–15. https://hdl.handle.net/10344/2151.

Fiebig, André. 2015. "Influence of context effects on sound quality assessments." *EuroNoise 2015*. Maastricht, Netherlands, May 31–June 3.

Flynn, James. 2022. "Full of sound and query, signifying something: Recent noise over acoustic trademarks." *International Lawyers Network*, September 28. https://www.ilnipinsider.com/2022/09/full-of-sound-and-query-signifying-something-recent-noise-over-acoustic-trademarks/.

Fontaine, Johnny R. J., Klaus R. Scherer, Etienne B. Roesch, and Phoebe C. Ellsworth. 2007. "The world of emotions is not two-dimensional." *Psychological Science* 18(12): 1050–1057.

Fortin, Pascal E., Elisabeth Sulmont, and Jeremy Cooperstock. 2019. "Detecting perception of smartphone notifications using skin conductance responses." *CHI 2019*

Proceedings of the 2019 CHI Conference on Human Factors in Computing Systems (1–12). Glasgow, May 4–9.

Forys, Brandon J., Emily Qi, Rebecca M. Todd, and Alan Kingstone. 2025. "Hear it here: Built environments predict ratings and descriptions of ambiguous sounds." *PLOS One* 20(1): e0316187. https://doi.org/10.1371/journal.pone.0316187.

Francis, C. D., P. Newman, B. D. Taff, C. White, C. A. Monz, M. Levenhagen, A. R. Petrelli, L. C. Abbott, J. Newton, S. Burson, and C. B. Cooper. 2017. "Acoustic environments matter: Synergistic benefits to humans and ecological communities." *Journal of Environmental Management* 203: 245–254.

Freed, Daniel J. 1990. "Auditory correlates of perceived mallet hardness for a set of recorded percussive sound events." *Journal of the Acoustical Society of America* 87: 311–322.

Frissen, Thomas. 2020. "Phantom cell phone signals/ringxiety." In *The International Encyclopedia of Media Psychology*, edited by Jan Van den Bulck, 1–8. New York: John Wiley & Sons. https://doi.org/10.1002/9781119011071.

Fritz, Jonathan B., Stephen V. David, Susanne Radtke-Schuller, Pingbo Yin, and Shihab A. Shamma. 2010. "Adaptive, behaviorally gated, persistent encoding of task-relevant auditory information in ferret frontal cortex." *Nature Neuroscience* 13(8): 1011–1019.

Fryer, Louise, Jonathan Freeman, and Linda Pring. 2014. "Touching words is not enough: How visual experience influences haptic–auditory associations in the 'Bouba–Kiki' effect." *Cognition* 132(2): 164–173.

Gacula, Maximo C., Jr. 1997. *Descriptive Sensory Analysis in Practice*. Trumbull, CT: Food & Nutrition Press.

Galmarini, Mara Virginia, R. J. Silva Paz, D. Enciso Choquehuanca, María Clara Zamora, and Bruno Mesz. 2021. "Impact of music on the dynamic perception of coffee and evoked emotions evaluated by temporal dominance of sensations (TDS) and emotions (TDE)." *Food Research International* 150: 110795.

Gardner, Bill. 2014. "From 'shrill' housewife to Downing Street: The changing voice of Margaret Thatcher." *Telegraph*. November 25. https://www.telegraph.co.uk/news/politics/11251919/From-shrill-housewife-to-Downing-Street-the-changing-voice-of-Margaret-Thatcher.html.

Garlin, Francine V., and Katherine Owen. 2006. "Setting the tone with the tune: A meta-analytic review of the effects of background music in retail settings." *Journal of Business Research* 59(8): 755–764.

Gaver, William W. 1989. "The SonicFinder: An interface that uses auditory icons." *Human–Computer Interaction* 4(1): 67–94.

Gaver, William W. 1993. "What in the world do we hear? An ecological approach to auditory event perception." *Ecological Psychology* 5: 1–29.

Gerrards-Hesse, Astrid, Kordelia Spies, and Friedrich W. Hesse. 1994. "Experimental inductions of emotional states and their effectiveness: A review." *British Journal of Psychology* 85(1): 55–78.

Ghani, Jody, Wolfgang Ellermeier, and Karin Zimmer. 2005. "A test battery measuring auditory capabilities of listening panels." *Forum Acusticum 2005*, Budapest, August 30–September 2.

Giacalone, Davide, Maciej Nitkiewicz, Samuel Moulin, Torstein Boðason, Jakob Lund Laugesen, and Søren Bech. 2017. "Sensory profiling of high-end loudspeakers using rapid methods—Part 2: Projective mapping with expert and naïve assessors." *Audio Engineering Society Convention 142*, Berlin, May 20–23.

Godfrey, A. Jonathan R., Malcolm Loudon, Meika Foster, and Joanne Hort. 2023. "Comparing traditional check-all-that-apply (CATA) and implicit response time go/no-go approaches for profiling consumer emotional response when tasting food." *Food Quality and Preference* 112: 105027.

Godøy, Rolf Inge, Egil Haga, and Alexander Refsum Jensenius. 2006. "Exploring music-related gestures by sound-tracing: A preliminary study." *Proceedings of the 2nd ConGAS International Symposium on Gesture Interfaces for Multimedia Systems*, Leeds, May 9–10.

Goodwin, Simon. N. 2009. "3-D sound for 3-D games—Beyond 5.1." *Audio Engineering Society Conference: 35th International Conference: Audio for Games*, London, February 11–13.

Google Design. 2019. Sound design and sonic brand (Google I/O'19). YouTube. https://www.youtube.com/watch?v=r6Nzqcotujc&feature=youtu.be.

Gopinath, Sumanth. 2021. "Beep: Listening to the digital watch." In *Media Infrastructures and the Politics of Digital Time: Essays on Hardwired Temporalities*, edited by A. Volmar and K. Stine. Amsterdam: Amsterdam University Press.

Gothelf, Jeff. 2013. *Lean UX: Applying Lean Principles to Improve User Experience*. Sebastopol, CA: O'Reilly Media.

Graakjaer, Nicolai Jørgensgaard, and Anders Bonde. 2018. "Non-musical sound branding: A conceptualization and research overview." *European Journal of Marketing* 52(7–8): 1505–1525.

Graham, Ruth. 2015. "This holiday season, talk to your baby boomer parents about their annoying cellphone noises." *Slate*, December 22. https://slate.com/culture/2015/12/how-to-talk-to-your-baby-boomer-parents-about-their-annoying-cellphone-noises.html.

Greever, Tom. 2020. *Articulating Design Decisions*. Sebastopol, CA: O'Reilly Media.

Grill, Thomas, Arthur Flexer, and Stuart Cunningham. 2011. "Identification of perceptual qualities in textural sounds using the repertory grid method." *Proceedings of the 6th Audio Mostly Conference*, Coimbra, Portugal, September 7–9, 67–74.

Grill-Spector, Kalanit, Richard Henson, and Alex Martin. 2006. "Repetition and the brain: Neural models of stimulus-specific effects." *Trends in Cognitive Sciences* 10(1): 14–23.

Groarke, Jenny M., Niamh MacCormac, Phoebe E. McKenna-Plumley, and Lisa Graham-Wisener. 2022. "Music listening was an emotional resource and social surrogate for older adults during the COVID-19 pandemic: A qualitative study." *Behaviour Change* 39: 168–179.

Guéguen, Nicolas, and Celine Jacob. 2002. "The influence of music on temporal perceptions in an on-hold waiting situation." *Psychology of Music* 30(2): 210–214.

Guest, Steve, Caroline Catmur, Donna Lloyd, and Charles Spence. 2002. "Audiotactile interactions in roughness perception." *Experimental Brain Research* 146: 161–171.

Guetta, Rachel, and Psyche Loui. 2017. "When music is salty: The crossmodal associations between sound and taste." *PLOS One* 12(3): e0173366.

Gura, David. 2010. "Noise from consumers prompts SunChips to return to traditional, quieter bags." *The Two Way, NPR*. October 6. https://www.npr.org/sections/thetwo-way/2010/10/06/130382547/noise-from-consumers-prompts-sunchips-to-go-back-to-traditional-packaging.

Gutton, Jean-Pierre. 2000. *Bruits et sons dans notre histoire: Essai sur la reconstitution du paysage sonore*. Paris: Presses Universitaires de France.

Hagtvedt, Henrik, and S. Adam Brasel. 2016. "Cross-modal communication: Sound frequency influences consumer responses to color lightness." *Journal of Marketing Research* 53: 551–562.

Hamari, Juho, Jonna Koivisto, and Harri Sarsa. 2014. "Does gamification work? A literature review of empirical studies on gamification." *47th Hawaii International Conference on System Sciences*, Waikoloa, January 6–9, 3025–3034.

Hansen, Liane. 2007. "Click, clack, ding! A look back at the typewriter." *NPR*. https://www.npr.org/transcripts/12900006.

Heider, Fritz, and Marianne Simmel. 1944. "An experimental study of apparent behavior." *American Journal of Psychology* 57(2): 243–259.

Heilman, Kenneth M. 1997. "The neurobiology of emotional experience." *Journal of Neuropsychiatry and Clinical Neurosciences* 9: 439–448.

Hermes, Kirsten, Tim Brookes, and Christopher Hummersone. 2015. "The influence of dumping bias on timbral clarity ratings." *139th Audio Engineering Society Convention*, New York, October 29–November 1.

Honing, Henkjan, and Olivia Ladinig. 2008. "The potential of the internet for music perception research: A comment on lab-based versus web-based studies." *Empirical Musicology Review* 3(1): 4–7.

Hsiao, Elvis. 2023. "Unboxing the delightful UX of Apple's boxes." *Fast Company*, July 3. https://www.fastcompany.com/90916642/unboxing-the-delightful-ux-of-apples-boxes.

Hu, Xiao, Fanjie Li, and Ruilun Liu. 2022. "Detecting music-induced emotion based on acoustic analysis and physiological sensing: A multimodal approach." *Applied Sciences* 12(18): 9354.

Hubert, Marco, Mirja Hubert, Arnd Florack, Marc Linzmajer, and Peter Kenning. 2013. "Neural correlates of impulsive buying tendencies during perception of product packaging." *Psychology & Marketing* 30(10): 861–873.

Hug, Daniel. 2017. "Hear me interact." In *Total Interaction: Theory and Practice of a New Paradigm for the Design Disciplines*, edited by Gerhard M. Buurman. Berlin: Birkhäuser.

Hug, Daniel, and Moritz Kemper. 2014. "From Foley to function: A pedagogical approach to sound design for novel interactions." *Journal of Sonic Studies* 6(1). https://www.researchcatalogue.net/view/237166/237167.

Hug, Daniel, and Simon Pfaff. 2019. "Bringing sound to interaction design: Challenges, opportunities, inspirations." In *Foundations in Sound Design for Embedded Media: A Multidisciplinary Approach*, edited by M. Filimowicz. New York: Routledge.

Hülsmeier, David, Lena Schell-Majoor, Jan Rennies, and Steven van de Par. 2014. "Perception of sound quality of product sounds: A subjective study using a semantic differential." *Proceedings of the International Congress on Noise Control Engineering*, Melbourne, November 16–19.

Hulusić, Vedad, Kurt Debattista, Vibhor Aggarwal, and Alan Chalmers. 2011. "Maintaining frame rate perception in interactive environments by exploiting audio-visual cross-modal interaction." *Visual Computers* 27(1): 57–66.

Hunter, Patrick G., and E. Glenn Schellenberg. 2011. "Interactive effects of personality and frequency of exposure on liking for music." *Personality and Individual Differences* 50(2): 175–179.

Hurlburt, Russell T., and Sarah A. Akhter. 2006. "The descriptive experience sampling method." *Phenomenology and the Cognitive Sciences* 5: 271–301.

Imschloss, Monika, and Christina Kuehnl. 2019. "Feel the music! Exploring the cross-modal correspondence between music and haptic perceptions of softness." *Journal of Retailing* 95(4): 158–169.

Jain, Dhruv, Angela Lin, Rose Guttman, Marcus Amalachandran, Aileen Zeng, Leah Findlater, and Jon Froehlich. 2019. "Exploring sound awareness in the home for people who are deaf or hard of hearing." *Proceedings of the 2019 CHI Conference on Human Factors in Computing Systems*, Glasgow, May 4–9.

Jakubowski, Kelly, Sebastian Finkel, Lauren Stewart, and Daniel Müllensiefen. 2017. "Dissecting an earworm: Melodic features and song popularity predict involuntary musical imagery." *Psychology of Aesthetics, Creativity, and the Arts* 11(2): 122–135.

Janowski, Maciej, and Maria Chełkowska-Zacharewicz. 2019. "What do we actually measure as music-induced emotions?" *Roczniki Psychologiczne* 4: 373–403.

Jansen, Reinier J., Elif Özcan, and René van Egmond. 2011. "PSST! Product Sound Sketching Tool." *Journal of the Audio Engineering Society* 59(6): 396–403.

Jeon, Myounghoon, Lee Ju-Hwan, Jason Sterkenburg, and Christopher Plummer. 2015. "Cultural differences in preference of auditory emoticons: USA and South Korea." *International Conference on Auditory Display*, Graz, Austria, July 8–10, https://digitalcommons.mtu.edu/cls-fp/60.

Jeong, Eunju, Hokyoung Ryu, Geonsang Jo, and Jaehyeok Kim. 2018. "Cognitive load changes during music listening and its implication in earcon design in public environments: An fNIRS study." *International Journal of Environmental Research and Public Health* 15(10): 2075.

Job, R. F. Soames. 1999. "Noise sensitivity as a factor influencing human reactions to noise." *Noise and Health* 1(3): 57–68.

Jousmäki, Veikko, and Riitta Hari. 1998. "Parchment-skin illusion: Sound-biased touch." *Current Biology* 8(6): 869–872.

Jurafsky, Dan. 2014. *The Language of Food: A Linguist Reads the Menu*. New York: W. W. Norton.

Juslin, Patrik N., Simon Liljeström, Daniel Västfjäll, Gonçalo Barradas, and Ana Silva. 2008. "An experience sampling study of emotional reactions to music: Listener, music, and situation." *Emotion* 8(5): 668–683. https://doi.org/10.1037/a0013505.

Juslin, Patrik N., and Daniel Västfjäll. 2008. "Emotional responses to music: The need to consider underlying mechanisms." *Behavioral and Brain Sciences* 31(5): 559–575.

Kassuba, Tanja, Mareike M. Menz, Brigitte Röder, and Hartwig R. Siebner. 2013. "Multisensory interactions between auditory and haptic object recognition." *Cerebral Cortex* 23(5): 1097–1107.

Kato, Takumi, Ryuji Yokote, Takashi Kondo, and Keizo Konoshi. 2020. "Effect of products' startup sound on repurchase intention." *International Journal of Japan Association for Management Systems* 12(1): 81–86.

Kellaris, James J., and Ronald C. Rice. 1993. "The influence of tempo, loudness, and gender of listener on responses to music." *Psychology & Marketing* 10(1): 15–29.

Keller, Kevin Lane. 1998. *Strategic Brand Management: Building, Measuring, and Managing Brand Equity*. Upper Saddle River, NJ: Prentice Hall.

Kemp, S. E., J. Hort, and T. Hollowood, eds. 2018. *Descriptive Analysis in Sensory Evaluation*. Hoboken, NJ: Wiley & Sons.

Kendall, Roger A., and Edward C. Carterette. 1993. "Verbal attributes of simultaneous wind instrument timbres: I. von Bismarck's adjectives." *Music Perception* 10(4): 445–467.

Khalfa, Stephanie, Simone Dalla Bella, Mathieu Roy, Isabelle Peretz, and Sonia J. Lupien. 2003. "Effects of relaxing music on salivary cortisol level after psychological stress." *Annals of the New York Academy of Sciences* 999(1): 374–376.

Khalfa, Stéphanie, Mathieu Roy, Pierre Rainville, Simone Dalla Bella, and Isabelle Peretz. 2008. "Role of tempo entrainment in psychophysiological differentiation of happy and sad music?" *International Journal of Psychophysiology* 68(1): 17–26.

Khosla, Proma. 2020. "Here's the story behind Netflix's iconic intro sound. A goat is involved." *Mashable*, August 5. https://mashable.com/article/netflix-ta-dum-sound-podcast-episode.

Kim, Ga-Won, and Myung Hwan Yun. 2018. "Understanding the impression of product sounds by integrating quantitative and qualitative findings." *International Journal of Industrial Ergonomics* 63: 98–109.

Kim, Nahye, Bonseung Koo, Jaehwa Yoon, and Kwangsu Cho. 2016. "Understanding the formation of user's first impression on an interface design from a neurophysiological perspective." *Proceedings of HCI Korea (HCIK '16)*, Jeongseon, South Korea, January 27–29. https://doi.org/10.17210/hcik.2016.01.139.

Kim, Seong-Hyeon, Dong-Chul Park, and Seok-Gwan Hong. 2012. "A study on regional and individual preference sound quality for luxury vehicle." *Proceedings of the Korean Society for Noise and Vibration Engineering Conference*, 364–369. https://www.koreascience.or.kr/article/CFKO201218139809689.pdf.

King, Silvia C., Herbert L. Meiselman, and B. Thomas Carr. 2013. "Measuring emotions associated with foods: Important elements of questionnaire and test design." *Food Quality and Preference* 28: 8–16.

Klanovicz, Cristiano Porto, Charles Spence, and Leandro Miletto Tonetto. 2022. "Product interaction sounds influence product personality." *Journal of Design Research* 20(2): 123–141.

Koelsch, Stefan, and Lutz Jäncke. 2015. "Music and the heart." *European Heart Journal* 36: 3043–3049. https://doi.org/10.1093/eurheartj/ehv430.

Köhler, Wolfgang. 1947. *Gestalt Psychology*. New York: Liveright.

Komatsu, T. 2005. "Can we assign attitudes to a computer based on its beeps?" *Proceedings of the Nineteenth International Joint Conference on Artificial Intelligence (IJCAI–05)*, Edinburgh, July 30–August 5.

Koshimori, Yuko. 2019. "Neurochemical responses to music." In *The Oxford Handbook of Music and the Brain*, edited by M. H. Thaut and D. A. Hodges. Oxford: Oxford University Press.

Kosslyn, Stephen M., Giorgio Ganis, and William L. Thompson. 2001. "Neural foundations of imagery." *Neuroscience* 2(9): 635–642.

Kreibig, Sylvia D. 2010. "Autonomic nervous system activity in emotion: A review." *Biological Psychology* 84(3): 394–421.

Krippendorff, Klaus. 2018. *Content Analysis: An Introduction to Its Methodology*. New York: Sage.

Kuckartz, Udo, and Stefan Rädiker. 2023. *Qualitative Content Analysis*. New York: Sage.

Kühn, Simone, Timothy R. Brick, Barbara C. N. Müller, and Jürgen Gallinat. 2014. "Is this car looking at you? How anthropomorphism predicts fusiform face area activation when seeing cars." *PLOS One* 9(12): e113885. https://doi.org/10.1371/journal.pone.0113885.

Kuliga, Saskia F., Tyler Thrash, Ruth Conroy Dalton, and Christoph Hölscher. 2015. "Virtual reality as an empirical research tool: Exploring user experience in a real building and a corresponding virtual model." *Computers, Environment and Urban Systems* 54: 363–375.

Kumar, Sukhbinder, Katharina von Kriegstein, Karl Friston, and Timothy D. Griffiths. 2012. "Features versus feelings: Dissociable representations of the acoustic features and valence of aversive sounds." *Journal of Neuroscience* 32(41): 14184.

Küssner, Mats B., Liila Taruffi, and Georgia A. Floridou, eds. 2023. *Music and Mental Imagery*. New York: Routledge

Kwa, Eng Keat, Soon Keng Cheong, Lin Kooi Ong, and Poh Foong Lee. 2024. "Development of audio-guided deep breathing and auditory go/no-go task on evaluating its impact on the wellness of young adults: A pilot study." *Biomedical Engineering/Biomedizinische Technik* 69(2): 141–150.

Kwon, Young-Sung, Jonghyun Lee, and Slgi Lee. 2022. "The impact of background music on film audience's attentional processes: Electroencephalography alpha-rhythm and event-related potential analyses." *Frontiers in Psychology* 13: 933497. https://doi.org/10.3389/fpsyg.2022.933497.

Kwon, Young-Sung, and Sung-Eun Lee. 2020. "Cognitive processing of sound effects in television sports broadcasting." *Journal of Radio & Audio Media* 27(1): 93–118. https://doi.org/10.1080/19376529.2018.1541899.

Laeng, Bruno, Sylvain Sirois, and Gustaf Gredebäck. 2012. "Pupillometry: A window to the preconscious?" *Perspectives on Psychological Science* 7(1): 18–27. https://doi.org/10.1177/1745691611427305.

Lahdelma, Imre, George Athanasopoulos, and Tuomas Eerola. 2021. "Sweetness is in the ear of the beholder: Chord preference across United Kingdom and Pakistani listeners." *Annals of the New York Academy of Sciences* 1502(1): 72–84.

Lahdelma, Imre, Tuomas Eerola, and James Armitage. 2022. "Is harmonicity a misnomer for cultural familiarity in consonance preferences?" *Frontiers in Psychology* 13: 802385. https://doi.org/10.3389/fpsyg.2022.802385.

Lamont, Alexandra. 2009. "Strong experiences of music in university students." *Proceedings of the 7th Triennial Conference of European Society for the Cognitive Sciences of Music*, Jyväskylä, Finland, August 12–16.

Lantos, Geoffrey P., and Lincoln G. Craton. 2012. "A model of consumer response to advertising music." *Journal of Consumer Marketing* 29(1): 22–42.

Lavie, Nilli, and Jan De Fockert. 2005. "The role of working memory in attentional capture." *Psychonomic Bulletin & Review* 12: 669–674.

Lechat, Bastien, Hannah Scott, Felix Decup, Kristy L. Hansen, Gorica Micic, Claire Dunbar, Tessa Liebich, Peter Catcheside, and Branko Zajamsek. 2022. "Environmental noise-induced cardiovascular responses during sleep." *Sleep 45*(3): zsab302.

Lederman, Susan J. 1979. "Auditory texture perception." *Perception* 8(1): 93–103.

LeDoux, Joseph E. 2000. "Emotion circuits in the brain." *Annual Review of Neuroscience* 23(1): 155–184.

Lee, Yi-Chieh, Wen-Chieh Lin, Jung-Tai King, Li-Wei Ko, Yu-Ting Huang, and Fu-Yin Cherng. 2014. "An EEG-based approach for evaluating audio notifications under ambient sounds." *Proceedings of the SIGCHI Conference on Human Factors in Computing Systems*, Toronto, April 26–May 1.

Lemaitre, Guillaume, Olivier Houix, Nicolas Misdariis, and Patrick Susini. 2010. "Listener expertise and sound identification influence the categorization of environmental sounds." *Journal of Experimental Psychology: Applied* 16(1): 16.

Lemaitre, Guillaume, Ali Jabbari, Olivier Houix, Nicolas Misdariis, and Patrick Susini. 2016. "Vocal imitations of basic auditory features." *Journal of the Acoustical Society of America* 139(1): 290–300.

Lemaitre, Guillaume, Hugo Scurto, Jules Françoise, Frédéric Bevilacqua, Olivier Houix, and Patrick Susini. 2017. "Rising tones and rustling noises: Metaphors in gestural depictions of sounds." *PLOS One* 12(7): e0181786.

Leman, Mark. 2008. *Embodied Music Cognition and Mediation Technology*. Cambridge, MA: MIT Press.

Lerma, Beatrice, Doriana Dal Palu, and Eleonora Buiatti. 2017. "The taste of the sound or the sound of the taste? How sounding packaging influences food perception." *3rd International Conference on Food Design*, virtual, February 17–18.

Levenson, Robert W., Paul Ekman, and Wallace V. Friesen. 1990. "Voluntary facial action generates emotion-specific autonomic nervous system activity." *Psychophysiology* 27(4): 363–384.

Liebetrau, Judith, Johannes Nowak, Thomas Sporer, Matthias Krause, Martin Rekitt, and Sebastian Schneider. 2013. "Paired comparison as a method for measuring emotions." *Audio Engineering Society Convention 135*, New York, October 17–20.

Lim, Juyun. 2011. "Hedonic scaling: A review of methods and theory." *Food Quality and Preference* 22(8): 733–747.

Lin, Meng-Hsien, Samantha N. N. Cross, William J. Jones, and Terry L. Childers. 2018. "Applying EEG in consumer neuroscience." *European Journal of Marketing* 52: 66–91. https://doi.org/10.1108/EJM-12-2016-0805.

Lipscomb, Scott D., and Mark Kerins. 2004. "An empirical investigation into the effect of presentation mode in the cinematic and music listening experience." *Proceedings of the 8th International Conference on Music Perception & Cognition*, Chicago, August 3–7.

Lorho, Gaetan, G. L. Ray, and Nick Zacharov. 2010. "eGauge: A measure of assessor expertise in audio quality evaluations." https://api.semanticscholar.org/CorpusID:54140463.

Loureiro, Sandra Maria Correia, João Guerreiro, and Arnold Japutra. 2021. "How escapism leads to behavioral intention in a virtual reality store with background music?" *Journal of Business Research* 134: 288–300.

Lowe, Michael L., Katherine E. Loveland, and Aradhna Krishna. 2019. "A quiet disquiet: Anxiety and risk avoidance due to nonconscious auditory priming." *Journal of Consumer Research* 46(1): 159–179.

Lundqvist, Lars-Olov, Fredrik Carlsson, Per Hilmersson, and Patrik N. Juslin. 2008. "Emotional responses to music: Experience, expression, and physiology." *Psychology of Music* 37(1): 61–90.

Macefield, Ritch. 2009. "How to specify the participant group size for usability studies: A practitioner's guide." *Journal of Usability Studies* 5(1): 34–45.

Madison, Guy, and Gunilla Schiölde. "Repeated listening increases the liking for music regardless of its complexity: Implications for the appreciation and aesthetics of music." *Frontiers in Neuroscience* 11 (2017): 147.

Magalhães, Mariana, António Coelho, Miguel Melo, and Maximino Bessa. 2024. "Measuring users' emotional responses in multisensory virtual reality: A systematic literature review." *Multimedia Tools and Applications* 83(14): 43377–43417.

Mandryk, Regan L., Kori M. Inkpen, and Thomas W. Calvert. 2006. "Using psychophysiological techniques to measure user experience with entertainment technologies." *Behaviour & Information Technology* 25(2): 250–262.

Mansell, James. 2017. "'A chamber of noise horrors': Sound, technology and the museum." *Science Museum Group Journal*, Spring. https://dx.doi.org/10.15180/170704.

Marcosson, Isaac Frederick. 1945. *Wherever Men Trade: The Romance of the Cash Register*. New York: Dodd, Mead.

Marentakis, Georgios. 2023. "Sketching in sound and interaction design." *Proceedings of Forum Acusticum 2023: 10th Convention of the European Acoustics Association*, Turin, Italy, September 11–15. https://dael.euracoustics.org/confs/fa2023/data/articles/001127.pdf.

Marsh, Stephanie. 2022. *User Research: Improve Product and Service Design and Enhance Your UX Research*. London: Kogan Page.

Mas-Herrero, Ernest, Alain Dagher, Marcel Farrés-Franch, and Robert J. Zatorre. 2021. "Unraveling the temporal dynamics of reward signals in music-induced pleasure with TMS." *Journal of Neuroscience* 41(17): 3889–3899.

Mattos, Evelio. 2022. "How packaging sounds is as important as it looks." *Unboxed Podcast*, June 22, episode 96. https://www.packagingunboxd.com/packaging-design-how-your-packaging-sounds/.

Max, Caroline, Andreas Widmann, Sonja A. Kotz, Erich Schröger, and Nicole Wetzel. 2015. "Distraction by emotional sounds: Disentangling arousal benefits and orienting costs." *Emotion* 15(4): 428–437.

McAdams, Stephen, Suzanne Winsberg, Sophie Donnadieu, Geert De Soete, and Jochen Krimphoff. 1995. "Perceptual scaling of synthesized timbres: Common dimensions, specificities, and latent subject classes." *Psychological Research* 58: 177–192.

McCann, Erin. 2019. "Innocent songs completely perverted by gruesome movie scenes." *Ranker*. October 30. https://www.ranker.com/list/innocent-songs-ruined-by-gruesome-movies/erin-mccann.

McCrie, Robert D. 1988. "The development of the U.S. security industry." *Annals of the American Academy of Political and Social Science* 498: 23–33.

McGee, Marilyn Rose, Phil Gray, and Stephen Brewster. 2001. "The effective combination of haptic and auditory textural information." *International Workshop on Haptic Human-Computer Interaction*, Glasgow, August 31–September 1.

McGregor, Iain. 2014. "Comparing designers' and listeners' experiences." *AI & Society* 29(4): 473–483. https://doi.org/10.1007/s00146-013-0489-4.

McGregor, Iain, Phil Turner, and David Benyon. 2014. "Using participatory visualization of soundscapes to compare designers' and listeners' experiences of sound designs." *Journal of Sonic Studies* 6(1). https://www.researchcatalogue.net/view/242648/242649.

McLuhan, Marshall, and Quentin Fiore. 2003. *The Medium Is the Massage: An Inventory of Effects*. Toronto: Penguin Canada.

McMullin, Elisabeth, Adrian Celestinos, and Allan Devantier. 2015. "Environments for evaluation: The development of two new rooms for subjective evaluation." *139th Audio Engineering Society Convention*, New York, October 29–November 1.

McPhetres, Jonathon, and Janis H. Zickfeld. 2022. "The physiological study of emotional piloerection: A systematic review and guide for future research." *International Journal of Psychophysiology* 179: 6–20.

Mehta, Ravi, Rui Zhu, and Amar Cheema. 2012. "Is noise always bad? Exploring the effects of ambient noise on creative cognition." *Journal of Consumer Research* 39(4): 784–799.

Mellander, Charlotta, Richard Florida, Peter J. Rentfrow, and Jeff Potter. 2018. "The geography of music preferences." *Journal of Cultural Economics* 42: 593–618.

Mesz, Bruno, Mariano Sigman, and Marcos A. Trevisan. 2012. "A composition algorithm based on crossmodal taste-music correspondences." *Frontiers in Human Neuroscience* 6: 71.

Michalakoudis, Ioannis, Peter Childs, Marco Aurisicchio, and James Harding. 2017. "Using functional analysis diagrams to improve product reliability and cost." *Advances in Mechanical Engineering* 9(1): 1–11. https://doi.org/10.1177/1687814016685223.

Miles, Matthew B., and A. Michael Huberman. 1994. *Qualitative Data Analysis*. Thousand Oaks, CA: Sage.

Miller, Darryl W., and Lawrence J. Marks. 1992. "Mental imagery and sound effects in radio commercials." *Journal of Advertising* 21(4): 83–93.

Milne, Alice E., Roberta Bianco, Katarina C. Poole, Sijia Zhao, Andrew J. Oxenham, Alexander J. Billig, and Maria Chait. 2020. "An online headphone screening test based on dichotic pitch." *Behavior Research Methods* 53: 1151–1562.

Minsky, Laurence, and Colleen Fahey. 2017. *Audio Branding: Using Sound to Build Your Brand*. London: Kogan Page.

Moerman, Jean Paul. 2004. "Loudness in TV sound." *Audio Engineering Society Convention 116*, Berlin, May 8–11.

Moesgaard, Frederik, Lasse Hulgaard, and Mads Bødker. 2020. "Involving users in sound design." *Design, User Experience, and Usability. Interaction Design: 9th International Conference*, Copenhagen, July 19–24.

Mok, Brittany A., Vibha Viswanathan, Agudemu Borjigin, Ravinderjit Singh, Homeira Kafi, and Hari M. Bharadwaj. 2023. "Web-based psychoacoustics: Hearing screening, infrastructure, and validation." *Behavior Research Methods* 56(3): 1433–1448. https://doi.org/10.3758/s13428-023-02101-9.

Monache, Stefano Delle, and Davide Rocchesso. 2016. "Cooperative sound design: A protocol analysis." *Proceedings of AudioMostly*, Norrköping, Sweden, October 4–6.

Monache, Stefano Delle, and Davide Rocchesso. 2019. "Exploring design cognition in voice-driven sound sketching and synthesis." In *Foundations in Sound Design for Embedded Media: A Multidisciplinary Approach*, edited by M. Filimowicz, 79–101. New York: Routledge.

Monfardini, Elisabetta, Jérôme Redouté, Fadila Hadj-Bouziane, Clément Hynaux, Jacques Fradin, Pascal Huguet, Nicolas Costes, and Martine Meunier. 2016. "Others' sheer presence boosts brain activity in the attention (but not the motivation) network." *Cerebral Cortex* 26(6): 2427–2439.

Moore, Dylan, Tobias Dahl, Paula Varela, Wendy Ju, Tormod Naes, and Ingunn Berget. 2019. "Unintended consonances: Methods to understand robot motor sound perception." *Proceedings of the 2019 CHI Conference on Human Factors in Computing Systems*, Glasgow, May 4–9.

Moreno, Roxana, and Richard E. Mayer. 2000. "A coherence effect in multimedia learning: The case for minimizing irrelevant sounds in the design of multimedia instructional messages." *Journal of Educational Psychology* 92(1): 117.

Morse, Margaret. 1998. *Virtualities: Television, Media Art and Cyberculture*. Bloomington: Indiana University Press.

Müllensiefen, Daniel. 2021. "Methods for testing the emotional effects of music in advertising and brand communication." In *Oxford Handbook of Music and Advertising*, edited by J. Deaville, 697–719. Oxford: Oxford University Press.

Naal-Ruiz, Norberto E., Luz M. Alonso-Valerdi, and David I. Ibarra-Zarate. 2021. "Different headphone models modulate differently alpha and theta brain oscillations when listening to the same sound." *43rd Annual International Conference of the IEEE Engineering in Medicine & Biology Society*, virtual conference, November 1–5, 1–4.

Nacke, Lennart Erik, Michael Kalyn, Calvin Lough, and Regan Lee Mandryk. 2011. "Biofeedback game design: Using direct and indirect physiological control to enhance game interaction." *CHI'11 Extended Abstracts on Human Factors in Computing Systems, CHI Conference on Human Factors in Computing Systems*, Vancouver, May 7–12, 103–112.

Nederhof, Anton J. 1985. "Methods of coping with social desirability bias: A review." *European Journal of Social Psychology* 15(3): 263–280.

Nees, Michael A., and Eliana Liebman. 2023. "Auditory icons, earcons, spearcons, and speech: A systematic review and meta-analysis of brief audio alerts in human-machine interfaces." *Auditory Perception & Cognition* 6(3–4): 300–329. https://doi.org/10.1080/25742442.2023.2219201.

Ngo, Mary Kim, Michelle L. Cadieux, Scott Sinnett, Salvador Soto-Faraco, and Charles Spence. 2011. "Reversing the Colavita visual dominance effect." *Experimental Brain Research* 214: 607–618.

Niedenthal, Paula M. 2007. "Embodying emotion." *Science* 316: 1002–1005.

Nielsen, Jakob, and Thomas K. Landauer. 1993. "A mathematical model of the finding of usability problems." *Proceedings of the INTERACT'93 and CHI'93 conference on Human Factors in Computing Systems*, Amsterdam, April 24–29, 206–213.

Nilsson, Ulrica. 2009. "Soothing music can increase oxytocin levels during bed rest after open-heart surgery: A randomized control trial." *Journal of Clinical Nursing* 18(15): 2153–2161.

Nishiguchi, Masayuki, Kazuki Ishimoto, Kanji Watanabe, Koji Abe, and Shouichi Takane. 2020. "Spatial auditory masking caused by phantom sound images." *Audio Engineering Society Convention 149*, online, October 27–30, eBrief 636.

Niu, Yunfang, Danli Wang, Ziwei Wang, Fan Sun, Kang Yue, and Nan Zheng. 2019. "User experience evaluation in virtual reality based on subjective feelings and physiological signals." *Electronic Imaging* 32: 1–11.

Noel, Jean-Paul, Hyeong-Dong Park, Isabella Pasqualini, Herve Lissek, Mark Wallace, Olaf Blanke, and Andrea Serino. 2018. "Audio-visual sensory deprivation degrades visuo-tactile peri-personal space." *Consciousness and Cognition* 61: 61–75.

Noetel, Michael, Shantell Griffith, Oscar Delaney, Nicola Rose Harris, Taren Sanders, Philip Parker, Borja del Pozo Cruz, and Chris Lonsdale. 2022. "Multimedia design for learning: An overview of reviews with meta-meta-analysis." *Review of Educational Research* 92(3): 413–454.

Norman, Donald. 2009. *The Design of Future Things*. New York: Basic Books.

North, Adrian C., Liam C. Mackenzie, Ruth M. Law, and David J. Hargreaves. 2004. "The effects of musical and voice 'fit' on responses to advertisements." *Journal of Applied Social Psychology* 34(8): 1675–1708.

Nosulenko, Valery, Etienne Parizet, and Elena Samoylenko. 2013. "The emotional component in perceived quality of noises produced by car engines." *International Journal of Vehicle Noise and Vibration* 9(1–2): 96–108.

Nosulenko, Valery N., and Elena S. Samoylenko. 1997. "Approche systémique de l'analyse des verbalisations dans le cadre de l'étude des processus perceptifs et cognitifs." *Social Sciences Information* 36(2): 223–261.

Novet, Jordan. 2021. "How software got so noisy, and why it's probably going to stay that way." *CNBC*, June 6. https://www.cnbc.com/2021/06/06/sound-design-for-software-why-software-makes-noise-and-how-its-made.html.

Nymoen, Kristian, Rolf Inge Godøy, Alexander Refsum Jensenius, and Jim Torresen. 2013. "Analyzing correspondence between sound objects and body motion." *ACM Transactions on Applied Perception* 10(2): 9:1–9:22.

Ohala, John J. 1983. "Cross-language use of pitch: An ethological view." *Phonetica* 40(1): 1–18.

Oliva, Manuel, and Andrey Anikin. 2018. "Pupil dilation reflects the time course of emotion recognition in human vocalizations." *Scientific Reports* 8(1): 4871.

Olsen, G. Douglas. 1995. "Creating the contrast: The influence of silence and background music on recall and attribute importance." *Journal of Advertising* 24(4): 29–44.

Otto, Norm, Scott Amman, Chris Eaton, and Scott Lake. 2001. "Guidelines for jury evaluations of automotive sounds." *Sound and Vibration*, April: 3015–3034.

Owczarzak, Mario. 2019. "Perceptions of audio notifications within a mobile messaging experience." Master's thesis, Purdue University.

Oxenham, Andrew J. 2018. "How we hear: The perception and neural coding of sound." *Annual Review of Psychology* 69(1): 27–50. https://doi.org/10.1146/annurev-psych-122216-011635.

Özcan, Elif, and René van Egmond. 2005. "Characterizing descriptions of product sounds." *Proceedings of the 11th International Conference on Auditory Display*, Limerick, Ireland, July 6–9, 55–60.

Özcan, Elif, and René van Egmond. 2010. "Audio-visual interactions in product sound design." *Human Vision and Electronic Imaging XV* 7527: 210–217.

Pannafino, James, and Patrick McNeil. 2017. *UX Methods: A Quick Guide to User Experience Research Methods*. Millersville, PA: CDUXP LLC.

Papadakis, Nikolaos M., Andreas Zantzas, Konstantinos Lafazanis, and Georgios E. Stavroulakis. 2022. "Influence of color on loudness perception of household appliances: Case of a coffee maker." *Designs* 6(6): 101.

Parasuraman, Raja, and Yang Jiang. 2012. "Individual differences in cognition, affect, and performance: Behavioral, neuroimaging, and molecular genetic approaches." *NeuroImage* 59(1): 70–82.

Parizet, Etienne, Nacer Hamzaoui, and Guillaume Sabatié. 2005. "Comparison of some listening test methods: A case study." *Acta Acustica United with Acustica* 91: 356–364.

Pedemonte, Marisa, Martín Testa, Marcela Díaz, and Diego Suarez-Bagnasco. 2014. "The impact of sound on electroencephalographic waves during sleep in patients suffering from tinnitus." *Sleep Science* 7(3): 143–151.

Peng, Z. Ellen, Sebastian Waz, Emily Buss, Yi Shen, Virginia Richards, Hari Bharadwaj, G. Christopher Stecker, Jordan A. Beim, Adam K. Bosen, Meredith D. Braza, Anna C. Diedesch, Claire M. Dorey, Andrew R. Dykstra, Frederick J. Gallun, Raymond L. Goldsworthy, Lincoln Gray, Eric C. Hoover, Antje Ihlefeld, Thomas Koelewijn, Judy G. Kopun, Juraj Mesik, Daniel E. Shub, and Jonathan H. Venezia. 2022. "Remote testing for psychological and physiological acoustics." *Journal of the Acoustical Society of America* 151: 3116–3128. https://doi.org/10.1121/10.0010422.

Pereira, Carlos Silva, João Teixeira, Patrícia Figueiredo, João Xavier, São Luís Castro, and Elvira Brattico. 2011. "Music and emotions in the brain: Familiarity matters." *PLOS One* 6(11): e27241.

Pirhonen, Antti, Kai Tuuri, Manne-Sakari Mustonen, and Emma Murphy. 2007. "Beyond clicks and beeps: In pursuit of an effective sound design methodology." *Haptic and Audio Interaction Design: Second International Workshop* LNCS 4813: 133–144.

Plourde-Kelly, Adam D., Kevin S. Saroka, and Blake T. Dotta. 2021. "The impact of emotionally valenced music on emotional state and EEG profile: Convergence of self-report and quantitative data." *Neuroscience Letters* 758: 136009.

Porter, Emily, and Patrick Kiernan. 2022. "State of the 'equal-loudness curves': A literature review." *Audio Engineering Society Convention 153*, Valencia, CA, October 19–20, Audio Engineering Society.

Poupis, Lauren Mayor, Dan Rubin, and Lama Lteif. 2021. "Turn up the volume if you're feeling lonely: The effect of mobile application sound on consumer outcomes." *Journal of Business Research* 126: 263–278.

Pradhan, Bikash K., Chinmaya R. Mishra, Maciej Jarzębski, J. Sivaraman, Sirsendu S. Ray, Satyapriya Mohanty, and Kunal Pal. 2022. "Investigating the effect of sound in horror clip on the cardiac electrophysiology of young adults using wavelet packet decomposition and machine learning classifiers." *Biomedical Engineering Advances* 3: 100037.

Presicce, Graziana, and Freya Bailes. 2019. "Engagement and visual imagery in music listening: An exploratory study." *Psychomusicology: Music, Mind and Brain* 29(2–3): 136–155.

Proverbio, Alice M., Luigi Manfrin, Laura A. Arcari, Francesco De Benedetto, Martina Gazzola, Matteo Guardamagna, Valentina Lozano Nasi, and Alberto Zan. 2015. "Non-expert listeners show decreased heart rate and increased blood pressure (fear bradycardia) in response to atonal music." *Frontiers in Psychology* 6: 1646.

Punch, Jerry, Antony Joseph, and Brad Rakerd. 2004. "Most comfortable and uncomfortable loudness levels." *American Journal of Audiology* 13(2): 144–157.

Rabinowitz, Neil C., Ben D. B. Willmore, Jan W. H. Schnupp, and Andrew J. King. 2011. "Contrast gain control in auditory cortex." *Neuron* 70(6): 1178–1191.

Radziun, Dominika, and H. Henrik Ehrsson. 2018. "Auditory cues influence the rubber-hand illusion." *Journal of Experimental Psychology: Human Perception and Performance* 44(7): 1012–1021.

Raja, Md Washim, Sandip Anand, and Ibha Kuma. 2020. "Multi-item scale construction to measure consumers' attitude toward advertising music." *Journal of Marketing Communications* 26(3): 314–327.

Ramage-Morin, Pamela L., Rex Banks, Dany Pineault, and Maha Atrach. 2019. "Unperceived hearing loss among Canadians aged 40 to 79." *Statistics Canada*. https://www150.statcan.gc.ca/n1/pub/82-003-x/2019008/article/00002-eng.htm.

Ramos, Liesl-Vivoni. 1993. "The effects of on-hold telephone music on the number of premature disconnections to a statewide protective services abuse hot line." *Journal of Music Therapy* 30(2): 119–129.

Reichardt, W., and H. Niese. 1970. "Choice of sound duration and silent intervals for test and comparison signals in the subjective measurement of loudness level." *Journal of the Acoustical Society of America* 47(4B): 1083–1090.

Remington, Anna, and Jake Fairnie. 2017. "A sound advantage: Increased auditory capacity in autism." *Cognition* 166: 459–465.

Rendall, Drew, Michael J. Owren, and Michael J. Ryan. 2009. "What do animal signals mean?" *Animal Behaviour* 78(2): 233–240.

Robb, James, Tom Garner, Karen Collins, and Lennart E. Nacke. 2017. "The impact of health-related sounds on player experience." *Simulation & Gaming* 48(3): 402–427.

Rochat, Philippe. 1998. "Self-perception and action in infancy." *Experimental Brain Research* 123(1–2): 102–109.

Rojas, David, Bill Kapralos, Andrew Hogue, Karen Collins, Lennart Nacke, Sayra Cristancho, Cristina Conati, and Adam Dubrowski. 2013. "Developing effective serious games: The effect of ambient auditory conditions on visual fidelity perception in stereoscopic 3D." *IEEE Transactions on Systems, Man, and Cybernetics* 43: 1572–1583.

Romagny, Sébastien, Tevy Sault, Clément Bouchet, Laure Thiebaut, Francesca Vincenzi, and David Morizet. 2024. "From noise to sound: Setting the base of packaging sound design for cosmetics by physical, sensory and cognitive characterization of lipstick closing sounds." *Food Quality and Preference* 113: 105058.

Rosi, Victor, Pablo Arias Sarah, Olivier Houix, Nicolas Misdariis, and Patrick Susini. 2023. "Shared mental representations underlie metaphorical sound concepts." *Scientific Reports* 13(1): 5180. https://doi.org/10.1038/s41598-023-32214-2.

Rosi, Victor, Aliette Ravillion, Olivier Houix, and Patrick Susini. 2022. "Best-worst scaling, an alternative method to assess perceptual sound qualities." *JASA Express Letters* 2(6): 064404. https://doi.org/10.1121/10.0011752.

Russell, James A. 1980. "A circumplex model of affect." *Journal of Personality and Social Psychology* 39(6): 1161.

Rychtáriková, Monika, Gerrit Vermeir, and Michael Vorländer. 2010. "Laboratory listening tests in building and room acoustics." *European Symposium Harmonization of European Sound Insulation Descriptors and Classification Standards*, Florence, Italy, December 10.

Salmi, Aleksi, Jie Li, and Katja Holtta-Otto. 2023. "Automatic facial expression analysis as a measure of user-designer empathy." *Journal of Mechanical Design* 145(3): 031403.

Samara, Anas, Leo Galway, Raymond Bond, and Hui Wang. 2019. "Affective state detection via facial expression analysis within a human–computer interaction context." *Journal of Ambient Intelligence and Humanized Computing* 10: 2175–2184.

Samermit, Patrawat, Jeremy Saal, and Nicolas Davidenko. 2019. "Cross-sensory stimuli modulate reactions to aversive sounds." *Multisensory Research* 32(3): 197–213. https://doi.org/10.1163/22134808-20191344.

Samoylenko, Elena, Stephen McAdams, and Valery Nosulenko. 1996. "Systematic analysis of verbalizations produced in comparing musical timbres." *International Journal of Psychology* 31(6): 255–278.

Schäfer, Katharina, and Tuomas Eerola. 2020. "How listening to music and engagement with other media provide a sense of belonging: An exploratory study of social surrogacy." *Psychology of Music* 48(2): 232–251.

Schafer, R. Murray. 1994. *The Soundscape: Our Sonic Environment and the Tuning of the World*. Rochester, VT: Destiny Books.

Schatz, Raimund, Sebastian Egger, and Kathrin Masuch. 2012. "The impact of test duration on user fatigue and reliability of subjective quality ratings." *Journal of the Audio Engineering Society* 60(1–2): 63–73.

Schellenberg, E. Glenn, Isabelle Peretz, and Sandrine Vieillard. 2008. "Liking for happy and sad sounding music: Effects of exposure." *Cognition and Emotion* 22: 218–237.

Scherer, Klaus R., Vera Shuman, Johnny R. J. Fontaine, and Cristina Soriano. 2013. "The GRID meets the wheel: Assessing emotional feeling via self-report." In *Components of Emotional Meaning: A Sourcebook*, edited by J. R. J. Fontaine, K. R. Scherer, and C. Soriano, 281–298. Oxford: Oxford University Press.

Schifferstein, Hendrik N. J., and Pieter M. A. Desmet. 2008. "Tools facilitating multisensory product design." *Design Journal* 11(2): 137–158.

Schlüter, Alexander, Matthias Waldkirch, Katrin Burmeister-Lamp, and Jan Auernhammer. 2021. "No second chance for a first impression: The role of aesthetics in early access video games." *International Journal of Innovation Management* 25(10): 2140002.

Schwartz, G. E., P. L. Fair, P. S. Greenberg, M. J. Friedman, and G. L. Klerman. 1974. "Facial EMG in the assessment of emotion." *Psychophysiology* 11(2): 237.

Sekuler, Robert, Allison B. Sekuler, and Renee Lau. 1997. "Sound alters visual motion perception." *Nature* 385(6614): 308. https://doi.org/10.1038/385308a0.

Senn, Olivier, Toni Amadeus Bechtold, Florian Hoesl, and Lorenz Kilchenmann. 2021. "Taste and familiarity affect the experience of groove in popular music." *Musicae Scientiae* 25(1): 45–66.

Serafine, Mary Louise, Robert G. Crowder, and Bruno H. Repp. 1984. "Integration of melody and text in memory for songs." *Cognition* 16: 285–303.

Shams, Ladan, Yukiyasu Kamitani, and Shinsuke Shimojo. 2000. "Illusions: What you see is what you hear." *Nature* 408(6814): 788. https://doi.org/10.1038/35048669.

Shedroff, Nathan, and Christopher Noessel. 2012. *Make It So: Interaction Design Lessons from Science Fiction*. New York: Rosenfeld.

Shin, Joongwon, and Anna S. Mattila. 2021. "Aww effect: Engaging consumers in 'non-cute' prosocial initiatives with cuteness." *Journal of Business Research* 126: 209–220.

Shintaro, Miyazaki. 2012. "Algorhythmic listening 1949–1962: Auditory practices of early mainframe computing." *AISB/IACAP World Congress*, Birmingham, July 2–6.

Sittler, Mareike C., Andrew J. Cooper, and Christian Montag. 2019. "Is empathy involved in our emotional response to music? The role of the PRL gene, empathy, and arousal in response to happy and sad music." *Psychomusicology: Music, Mind, and Brain* 29(1): 10–21.

Śliwińska-Kowalska, Mariola, and Kamil Zaborowski. 2017. "WHO environmental noise guidelines for the European region: A systematic review on environmental noise and permanent hearing loss and tinnitus." *International Journal of Environmental Research in Public Health* 14(10): 1139. https://doi.org/10.3390/ijerph14101139.

Söderqvist, Thomas, and Adam Bencard. 2010. "Do things talk?" In *The Exhibition as Product and Generator of Scholarship*, edited by S. Lehmann-Brauns, C. Sichau, and H. Trischler, 92–102. Berlin: Max Planck Institute for the History of Science.

Sorokowski, Piotr, David Puts, Janie Johnson, Olga Żółkiewicz, Anna Oleszkiewicz, Agnieszka Sorokowska, Marta Kowal, Barbara Borkowska, and Katarzyna Pisanski. 2019. "Voice of authority: Professionals lower their vocal frequencies when giving expert advice." *Journal of Nonverbal Behavior* 43: 257–269.

Sörqvist, Patrik, Stefan Stenfelt, and Jerker Rönnberg. 2012. "Working memory capacity and visual-verbal cognitive load modulate auditory-sensory gating in the brainstem: Toward a unified view of attention." *Journal of Cognitive Neuroscience* 24(11): 2147–2154.

Soto-Sanfiel, María T., Bradley C. Freeman, and Ariadna Angulo-Brunet. 2023. "Radio art: Mental images and appreciation." *International Journal of Listening* 37(3): 199–211.

Sova, Deborah Hinderer, and Jakob Nielsen. n.d. "234 tips and tricks for recruiting users as participants in usability studies." https://media.nngroup.com/media/reports/free/How_To_Recruit_Participants_for_Usability_Studies.pdf.

Spence, Charles. 2015. "Eating with our ears: Assessing the importance of the sounds of consumption on our perception and enjoyment of multisensory flavour experiences." *Flavour* 4(3): 1–14. https://flavourjournal.biomedcentral.com/articles/10.1186/2044-7248-4-3.

Spence, Charles, and Nicola Di Stefano. 2022. "Coloured hearing, colour music, colour organs, and the search for perceptually meaningful correspondences between colour and sound." *i-Perception* 13(3): 1–42.

Spence, Charles, and Qian Wang. 2017. "Assessing the impact of closure type on wine ratings and mood." *Beverages* 3(4): 52.

Spence, Charles, and Massimiliano Zampini. 2006. "Auditory contributions to multisensory product perception." *Acta Acustica United with Acustica* 92(6): 1009–1025.

Spence, Charles, and Massimiliano Zampini. 2007. "Affective design: Modulating the pleasantness and forcefulness of aerosol sprays by manipulating aerosol spraying sounds." *CoDesign* 3: 109–123.

Stanton, Tasha R., and Charles Spence. 2020. "The influence of auditory cues on bodily and movement perception." *Frontiers in Psychology* 10: 3001.

Steffens, Jochen, Franz Müller, Melanie Schulz, and Samuel Gibson. 2020. "The effect of inattention and cognitive load on unpleasantness judgments of environmental sounds." *Applied Acoustics* 164: 107278.

Steiner, Paul. 2025. "Sound marketing." In *Quick Guide Sound Marketing*. Wiesbaden: Springer Gabler. https://doi.org/10.1007/978-3-658-46056-3_4.

Strohmeier, Dominik, Satu Jumisko-Pyykkö, and Kristina Kunze. 2010. "Open profiling of quality: A mixed method approach to understanding multimodal quality perception." *Advances in Multimedia* 2010: 658980. https://doi.org/10.1155/2010/658980.

Stupacher, Jan, Michael J. Hove, and Petr Janata. 2016. "Audio features underlying perceived groove and sensorimotor synchronization in music." *Music Perception* 33(5): 571–589. https://doi.org/10.1525/mp.

Sunaga, Tsutomu. 2018. "How the sound frequency of background music influences consumers' perceptions and decision making." *Psychology & Marketing* 35: 253–267.

Susanto, Yosephine, Andrew G. Livingstone, Bee Chin Ng, and Erik Cambria. 2020. "The hourglass model revisited." *IEEE Intelligent Systems* 35(5): 96–102.

Susini, Patrick. 2000. "Psychophysical validation of a proprioceptive device by cross-modal matching of loudness." *Acta Acustica* 86(3): 515–525.

Susini, Patrick, Olivier Houix, Nicolas Misdariis, Bennett Smith, and Sabine Langlois. 2009. "Instruction's effect on semantic scale ratings of interior car sounds." *Applied Acoustics* 70(3): 389–403.

Susini, Patrick, Guillaume Lemaitre, and Stephen McAdams. 2012. "Psychological measurement for sound description and evaluation." In *Measurements with Persons: Theory, Methods, and Implementation Areas*, edited by B. Berglund, G. B. Rossi, J. T. Townsend, and L. R. Pendrill, 227–253. New York: Psychology Press.

Taffese, Tewodros. 2017. "A review of using EEG and EMG psychophysiological measurements in user experience research." Master's thesis, University of Oulu.

Tagg, Philip. 2004. "Gestural interconversion and connotative precision." https://www.tagg.org/articles/xpdfs/filminternat0412.pdf.

Tagg, Philip. 2005. "Musical microcosms and the political economy." Presentation for the IASPM conference, Rome. https://www.tagg.org/articles/Roma0507.html.

Tagg, Philip. 2008. "Intel inside analysis." YouTube. https://www.youtube.com/watch?v=p5ZsmHHmDGA.

Tagg, Philip. 2013. *Music's Meanings*. New York: Mass Media Music Scholars' Press.

Tagg, Philip. n.d. "Glossary of terms, neologisms, etc. used in writings." https://tagg.org/articles/ptgloss.html.

Tajadura-Jiménez, Ana, Laura Crucianelli, Rebecca Zheng, Chloe Cheng, Judith Ley-Flores, Mercedes Borda-Más, Nadia Bianchi-Berthouze, and Aikaterini Fotopoulou. 2022. "Body weight distortions in an auditory-driven body illusion in subclinical and clinical eating disorders." *Scientific Reports* 12(1): 20031.

Tajadura-Jiménez, Ana, Aleksander Väljamäe, Frédéric Bevilacqua, and Nadia Bianchi-Berthouze. 2017. "Principles for designing body-centered auditory feedback." In *The*

Wiley Handbook of Human Computer Interaction, edited by K. L. Norman and J. Kirakowski, 371–403. Hoboken, NJ: Wiley.

Tajadura-Jiménez, Ana, Aleksander Väljamäe, Iwaki Toshima, Toshitaka Kimura, Manos Tsakiris, and Norimichi Kitagawa. 2012. "Action sounds recalibrate perceived tactile distance." *Current Biology* 22(13): R516–R517.

Tan, Yi Ting, Isabelle Peretz, Gary E. McPherson, and Sarah J. Wilson. 2021. "Establishing the reliability and validity of web-based singing research." *Music Perception* 38(4): 386–405.

Tang, Huimin, Boon Giin Lee, Dave Towey, Kaiyi Chen, Yichu Fang, Runzhou Zhang, and Matthew Pike. 2024. "The audience effect: Do observations change outcomes in HCI studies?" *2024 IEEE 48th Annual Computers, Software, and Applications Conference (COMPSAC)*, Osaka, Japan, July 2–4, 611–620. https://doi.org/10.1109/COMPSAC61105.2024.00088.

Tardieu, Julien, Patrick Susini, Franck Poisson, Hiroshi Kawakami, and Stephen McAdams. 2009. "The design and evaluation of an auditory way-finding system in a train station." *Applied Acoustics* 70(9): 1183–1193.

Taruffi, Liila, and Mats B. Küssner. 2019. "A review of music-evoked visual mental imagery: Conceptual issues, relation to emotion, and functional outcome." *Psychomusicology: Music, Mind, and Brain* 29(2–3): 62–74.

Teixeira, Thales, Rosalind Picard, and Rana El Kaliouby. 2014. "Why, when, and how much to entertain consumers in advertisements? A web-based facial tracking field study." *Marketing Science* 33(6): 809–827.

Think with Google. n.d. "The magic behind unboxing on YouTube." https://www.thinkwithgoogle.com/_qs/documents/119/youtube-insights-stats-data-trends-vol7_articles.pdf.

Thoma, Myriam V., Roberto La Marca, Rebecca Brönnimann, Linda Finkel, Ulrike Ehlert, and Urs M. Nater. 2013. "The effect of music on the human stress response." *PLOS One* 8(8): e70156. https://doi.org/10.1371/journal.pone.0070156.

Todd, Neil. 2001. "Evidence for a behavioral significance of saccular acoustic sensitivity in humans." *Journal of the Acoustical Society of America* 110(1): 380–390.

Tonetto, Leandro Miletto, Cristiano Porto Klanovicz, and Charles Spence. 2014. "Modifying action sounds influences people's emotional responses and bodily sensations." *I-Perception* 5(2): 153–163.

Travis, David, and Philip Hodgson. 2019. *Think like a UX Researcher: How to Observe Users, Influence Design and Shape Business Strategy*. Boca Raton, FL: CRC Press.

Triantos, Alexandros, Emmanuella Plakoyiannaki, Evaggelia Outra, and Nikolaos Petridis. 2016. "Anthropomorphic packaging: Is there life on Mars?" *European Journal of Marketing* 50(1–2): 260–275.

Trompeta, Maria-Angeliki, Kalipso Karantinou, Christos Koritos, and Tammo H. A. Bijmolt. 2022. "A meta-analysis of the effects of music in tourism and hospitality settings." *Journal of Business Research* 138: 130–145.

Tsow, Francis, Anupam Kumar, S. M. Hadi Hosseini, and Audrey Bowden. 2021. "A low-cost, wearable, do-it-yourself functional near-infrared spectroscopy (DIY-fNIRS) headband." *HardwareX* 10: e00204.

Ulanovsky, Nachum, Liora Las, and Israel Nelken. 2003. "Processing of low-probability sounds by cortical neurons." *Nature Neuroscience* 6(4): 391–398.

van der Zwaag, Marjolein D., Joyce H. D. M. Westerink, and Egon L. Van den Broek. 2011. "Emotional and psychophysiological responses to tempo, mode, and percussiveness." *Musicae Scientiae* 15(2): 250–269.

Vanstone, Ashley D., Lola L. Cuddy, Jacalyn M. Duffin, and Elizabeth Alexander. 2009. "Exceptional preservation of memory for tunes and lyrics: Case studies of amusia, profound deafness, and Alzheimer's disease." *Annals of the New York Academy of Sciences* 1169: 291–294.

Västfjäll, Daniel. 2003. "Emotional reactions to product sounds." *Proceedings of Journée Human Engineering*, Paris, March 20–June 12.

Västfjäll, Daniel. 2004. "Contextual influences on sound quality evaluation." *Acta Acustica United with Acustica* 90: 1029–1036.

Velasco, Carlos, Guido Corradi, and Kosuke Motoki. 2023. "Harmony of senses: Exploring the impact of sound aesthetic features on taste imagery." *Food Quality and Preference* 111: 104992.

Velasco, Carlos, and Charles Spence, eds. 2019. *Multisensory Packaging: Designing New Product Experiences*. Cham: Springer International.

Velasco, Franklin, Zhiyong Yang, and Narayanan Janakiraman. 2021. "A meta-analytic investigation of consumer response to anthropomorphic appeals: The roles of product type and uncertainty avoidance." *Journal of Business Research* 131: 735–746.

Vidal, Leticia, Gastón Ares, and Ana Giménez. 2013. "Projective techniques to uncover consumer perception: Application of three methodologies to ready-to-eat salads." *Food Quality and Preference* 28(1): 1–7.

Viollon, Stéphanie, Catherine Lavandier, and Carolyn Drake. 2002. "Influence of visual setting on sound ratings in an urban environment." *Applied Acoustics* 63(5): 493–511.

Visell, Yon. 2009. "Tactile sensory substitution: Models for enaction in HCI." *Interacting with Computers* 21(1–2): 38–53.

Voit, Alexandra, Sven Mayer, Valentin Schwind, and Niels Henze. 2019. "Online, VR, AR, lab, and in-situ: Comparison of research methods to evaluate smart artifacts."

Proceedings of the 2019 ACM CHI Conference on Human Factors in Computing Systems, Glasgow, May 4–9.

Volk, Christer, Søren Bech, Torben H. Pedersen, and Flemming Christensen. 2015. "Five aspects of maximizing objectivity from perceptual evaluations of loudspeakers: A literature study." *138th Audio Engineering Society Convention*, Warsaw, May 7–10.

Vowels, M. J., and Russell Mason. 2020. "Comparison of pairwise dissimilarity and projective mapping tasks with auditory stimuli." *Journal of the Audio Engineering Society* 68(9): 638–648.

Vroomen, Jean, Paul Bertelson, and Beatrice De Gelder. 2001. "The ventriloquist effect does not depend on the direction of automatic visual attention." *Perception & Psychophysics* 63(4): 651–659.

Walla, Peter, Gerhard Brenner, and Monika Koller. 2011. "Objective measures of emotion related to brand attitude: A new way to quantify emotion-related aspects relevant to marketing." *PLOS One* 6(11): e26782. https://doi.org/10.1371/journal.pone.0026782.

Walton, Tim, and Michael Evans. 2018. "The role of human influence factors on overall listening experience." *Quality and User Experience* 3(1): 1–16.

Walton, Tim, Michael Evans, Frank Melchior, and David Kirk. 2017. "Combining preference ratings with sensory profiling for the comparison of audio reproduction systems." *142nd Audio Engineering Society Convention*, Berlin, May 20–23.

Wan, Echo Wen, and Rocky Peng Chen. 2021. "Anthropomorphism and object attachment." *Current Opinion in Psychology* 39: 88–93.

Wan, Jing, and Pankaj Aggarwal. 2015. "Befriending Mr. Clean: The role of anthropomorphism in consumer-brand relationships." In *Strong Brands, Strong Relationships*, edited by S. Fournier, M. Breazeale, and J. Avery, 119–134. New York: Routledge. https://doi.org/10.4324/9781315767079-20.

Wang, Cai, Soo Yeon Kim, Yein Song, Sungho Kim, Minsik Choi, Donghoon Seu, and Myung Hwan Yun. 2022. "Text mining for exploring UX issues of qualitative think aloud data on EV sound." *2022 IEEE International Conference on Industrial Engineering and Engineering Management*, Kuala Lumpur, December 7–10, 200–203.

Wang, Kainan S., and Mauricio R. Delgado. 2021. "The protective effects of perceived control during repeated exposure to aversive stimuli." *Frontiers in Neuroscience* 15: 625816. https://doi.org/10.3389/fnins.2021.625816.

Wang, Lei, Da Qian, and Ou Li. 2020. "The cross-modal interaction between sound frequency and color saturation on consumer's product size perception, preference, and purchase." *Psychology and Marketing* 37(7): 876–899.

Wankling, Matthew, Bruno Fazenda, and William J. Davies. 2012. "The assessment of low-frequency room acoustic parameters using descriptive analysis." *Journal of the Audio Engineering Society* 60(5): 325–337.

Watanabe, Keita, and M. Yasumura. 2003. "Realizable user interface: The information realization by using cursor." *Proceedings of Human Interface Symposium*, Crete, June 22–27, 541–544.

Wetzel, Nicole, Andreas Widmann, and Florian Scharf. 2021. "Distraction of attention by novel sounds in children declines fast." *Nature Scientific Reports* 11: 5308.

Wharton, Alexander, and Karen Collins. 2011. "Subjective measures of the influence of music customization on the video game play experience: A pilot study." *Game Studies* 11(2): https://gamestudies.org/1102/articles/wharton_collins.

Williams, Duncan, Chia Yu Wu, Victoria J. Hodge, Damian Murphy, and Peter I. Cowling. 2015. "A psychometric evaluation of emotional responses to horror music." *Audio Engineering Society: 146th International Pro Audio Convention*, York, March 21–23.

Williams, Zachary J., Jason L. He, Carissa J. Cascio, and Tiffany G. Woynaroski. 2021. "A review of decreased sound tolerance in autism: Definitions, phenomenology, and potential mechanisms." *Neuroscience & Biobehavioral Reviews* 121: 1–17.

Willmore, Ben D. B., and Andrew J. King. 2023. "Adaptation in auditory processing." *Physiological Reviews* 103(2): 1025–1058.

Willson, Beckles. 1933. "The Anti-Noise League." *British Medical Journal* 2(3799): 797–798.

Witchel, Harry J., Lisa Lavia, Carina E. I. Westling, Aoife Healy, Robert Needham, and Nachiappan Chockalingam. 2013. "Using body language indicators for assessing the effects of soundscape quality on individuals." *Proceedings of the AIA-DAGA Conference on Acoustics*, Merano, Italy, March 18–21.

Wolkomir, Richard. 1996. "Decibel by decibel, reducing the din to a very dull roar." *Smithsonian Magazine*, 56–65. https://acoustics.org/pressroom/httpdocs/swa9602.html.

Woods, Kevin J. P., Max H. Siegel, James Traer, and Josh H. McDermott. 2017. "Headphone screening to facilitate web-based auditory experiments." *Attention, Perception, & Psychophysics* 79(7): 2064–2072. https://doi.org/10.3758/s13414-017-1361-2.

Woolrych, Alan, and Gilbert Cockton. 2001. "Why and when five test users aren't enough." In *Proceedings of the Proceedings of IHM-HCI 2001 Conference*, edited by J. Vanderdonckt, A. Blandford, and A. Derycke. London: Cépaduès Editions, 105–108.

Wu, Jianhui, Xiaoqin Mai, Chetwyn C. H. Chan, Yaqin Zheng, and Yuejia Luo. 2006. "Event-related potentials during mental imagery of animal sounds." *Psychophysiology* 43: 592–597.

Wycisk, Yves, Reinhard Kopiez, Jakob Bergner, Kilian Sander, Stephan Preihs, Jürgen Peissig, and Friedrich Platz. 2023. "The Headphone and Loudspeaker Test—Part I: Suggestions for controlling characteristics of playback devices in internet experiments." *Behavior Research Methods* 55: 1094–1107.

Yalch, Richard F. 1991. "Memory in a jingle jungle: Music as a mnemonic device in communicating advertising slogans." *Journal of Applied Psychology* 76(2): 268–275.

Yalch, Richard F., and Eric R. Spangenberg. 2000. "The effects of music in a retail setting on real and perceived shopping times." *Journal of Business Research* 49(2): 139–147.

Yanagisawa, Hideyoshi, Atsushi Kataoka, Tamotsu Murakami, Koichi Ohtomi, and Rika Hosaka. 2009. "Extraction of latent emotional factors by analyzing human sensitivity towards unexplored design: Application to product sound design." *Proceedings of the International Conference on Engineering Design*, Stanford, CA, August 24–27.

Yau, Jeffrey M., Jonathon B. Olenczak, John F. Dammann, and Sliman J. Bensmaia. 2009. "Temporal frequency channels linked across audition and touch." *Current Biology* 19(7): 561–566.

Yu, Lei, and Jian Kang. 2010. "Factors influencing the sound preference in urban open spaces." *Applied Acoustics* 71(7): 622–633.

Yuan, Lingyao, and Alan Dennis. 2017. "Interacting like humans? Understanding the effect of anthropomorphism on consumer's willingness to pay in online auctions." *Proceedings of the 50th Hawaii International Conference on System Sciences*, Waikoloa, January 4–7.

Yuan, Lingyao, and Alan R. Dennis. 2019. "Acting like humans? Anthropomorphism and consumer's willingness to pay in electronic commerce." *Journal of Management Information Systems* 36(2): 450–477.

Zacharov, Nick. 2020. *Sensory Evaluation of Sound*. Boca Raton, FL: CRC Press.

Zacharov, Nick, and Kalle Koivuniemi. 2001. "Audio descriptive analysis and mapping of spatial sound displays." *Proceedings of the 2001 International Conference on Auditory Display*, Espoo, Finland, July 29–August 1.

Zajonc, Robert. B. 1968. "Attitudinal effects of mere exposure." *Journal of Personality and Social Psychology* 9(2): 1–27.

Zaki, Tarannum, and Muhammad Nazrul Islam. 2021. "Neurological and physiological measures to evaluate the usability and user-experience (UX) of information systems: A systematic literature review." *Computer Science Review* 40: 100375.

Zentner, Marcel, and Tuomas Eerola. 2010. "Self-report measures and models." In *Handbook of Music and Emotion*, edited by J. A. Sloboda. Oxford: Oxford University Press.

Zentner, Marcel, Didier Grandjean, and Klaus R. Scherer. 2008. "Emotions evoked by the sound of music: Characterization, classification, and measurement." *Emotion* 8(4): 494–521.

Zhang, Le-kai, Shou-qian Sun, Bai-xi Xing, Rui-ming Luo, and Ke-jun Zhang. 2019. "Using psychophysiological measures to recognize personal music emotional experience." *Frontiers of Information Technology & Electronic Engineering* 20: 964–974.

Zhang, Shiyi, Yan Liu, Fanghao Song, Dong Yu, Zhenming Bo, and Zihan Zhang. 2023. "The effect of audiovisual spatial design on user experience of bare-hand interaction in VR." *International Journal of Human–Computer Interaction* 40(11): 2796–2807. https://doi.org/10.1080/10447318.2023.2171761.

Zhang, Ziyi, Shuofei Zhu, Jaron Mink, Aiping Xiong, Linhai Song, and Gang Wang. 2022. "Beyond bot detection: Combating fraudulent online survey takers." *Proceedings of the ACM Web Conference 2022*, Lyon, France, April 25–29.

Zhao, Sijia, Christopher A. Brown, Lori L. Holt, and Frederic Dick. 2022. "Robust and efficient online auditory psychophysics." *Trends in Hearing* 26: 1–24.

Zhelezoglo, Karina N., Robbie J. Hanson, Caio F. Miguel, and Karen M. Lionello-DeNolf. 2021. "The establishment of auditory–visual equivalence classes with a go/no-go successive matching-to-sample procedure." *Journal of the Experimental Analysis of Behavior* 115: 421–438.

Zimmer, Karin, and Wolfgang Ellermeier. 1999. "Psychometric properties of four measures of noise sensitivity: A comparison." *Journal of Environmental Psychology* 19(3): 295–302.

Zorea, Marine, and Katsuhiko Kushi. 2023. "Audible imagery: Creative contemplations on the sounds of home." *Proceedings of the Creativity & Cognition '23*, virtual, June 19–21.

Media References

Bluey. 2020. Season 2, episode 28, "Movies," directed by Joe Brumm, aired on October 27. ABC Kids.

de Jongh, Adriaan, and Céline Veltman. 2017. *Hidden Folks*. Steam.

Eguchi, Katsuya, and Hisashi Nogami. 2001. *Animal Crossing*. Nintendo GameBoy.

Harris, Peter, dir. 1981. *The Muppets Go to the Movies*. Marble Arch Productions.

Kojima, Hideo. 1987. *Metal Gear Solid*. Nintendo Entertainment System, Konami.

Lucas, George, dir. 1977. *Star Wars*. Lucasfilm.

Miyamoto, Shigeru. 1985. *Super Mario Bros.* Nintendo Entertainment System.

Miyamoto, Shigeru, and Takashi Tezuka. 1986. *The Legend of Zelda*. Nintendo Entertainment System.

Monty Python's Flying Circus. 1969. Season 1, episode 6, "It's the Arts," directed by Ian MacNaughton, November 23. BBC.

Robot Chicken. 2006. Season 2, episode 23, "Easter Basket," directed by Tom Root, April 16. Cartoon Network.

Sanders, Chris, and Dean DeBlois, dirs. 2010. *How to Train Your Dragon*. DreamWorks.

Silverman, David, dir. 2007. *The Simpsons Movie*. 20th Century Fox.

Siodmak, Curt, and Herbert L. Strock, dirs. 1953. *The Magnetic Monster*. Ivan Tors Productions.

Spielberg, Steven, dir. 1975. *Jaws*. Universal Pictures.

Spielberg, Steven, dir. 2002. *Minority Report*. 20th Century Fox.

Stanton, Andrew, dir. 2008. *Wall-E*. Pixar Animation Studios.

Strock, Herbert L., dir. 1954. *Gog*. Ivan Tors Productions.

Index

Publisher contact:
The MIT Press
Massachusetts Institute of Technology
77 Massachusetts Avenue, Cambridge, MA 02139
mitpress.mit.edu

EU Authorised Representative:
Easy Access System Europe, Mustamäe tee 50,
10621 Tallinn, Estonia
gpsr.requests@easproject.com

Printed by Integrated Books International,
United States of America